Andrew Mitchell is a naturalist and television producer. After studying zoology at Bristol University,
he joined the Scientific Exploration Society in London and organized expeditions to many parts of the
world. He specialized in the study of primates and,
through Operations Drake and Raleigh, pioneered
the use of lightweight aerial walkways for exploring
the rainforest treetops in Panama, Papua New
Guinea, Indonesia and Costa Rica. Rainforests drew
him once again to the high islands of the South
Pacific, through which he travelled in 1986 and 1987
gathering material for this book.

His career in television – as producer, writer and
broadcaster – has included work on *The Amateur
Naturalist* with Gerald Durrell, *The Living Isles*
with Julian Pettifer, and the BBC's *Horizon* and
Tomorrow's World. He also worked on the feature
film *Greystoke: The Legend of Tarzan*, and co-pre-
sented the Channel 4 series *Odyssey*.

Andrew Mitchell lives in London. He is a fellow of
the Royal Geographical Society, a council member
of the Scientific Exploration Society, and deputy
Director of Earthwatch Europe. *A Fragile Paradise* is
his sixth book.

Books by Andrew Mitchell

Voyage of Discovery
The Young Naturalist
Reaching the Rainforest Roof
Rainforest Wildlife
The Enchanted Canopy

A FRAGILE PARADISE

Nature and Man in the Pacific

ANDREW MITCHELL

FONTANA/Collins

First published by William Collins 1989
First issued in Fontana Paperbacks 1990
Copyright © Andrew Mitchell 1989

Printed and bound in Great Britain by
Butler & Tanner Ltd, Frome and London

To John Gibbons

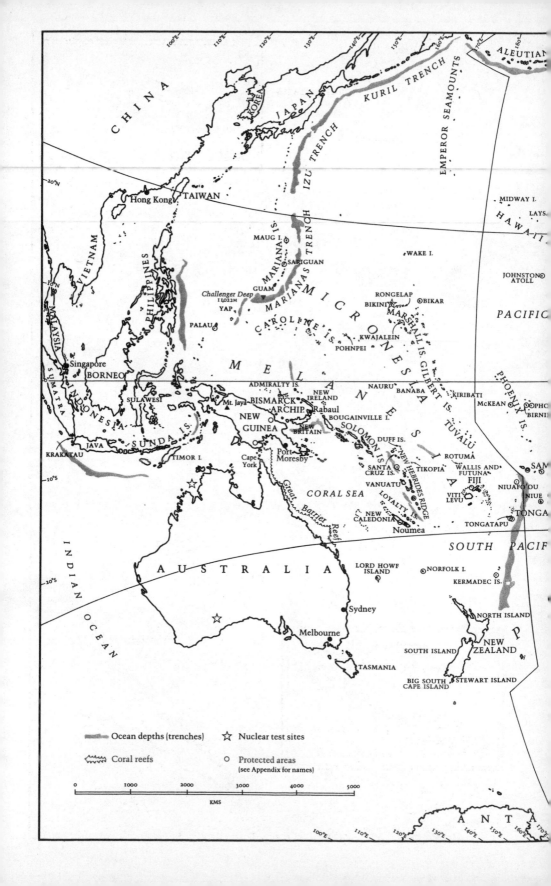

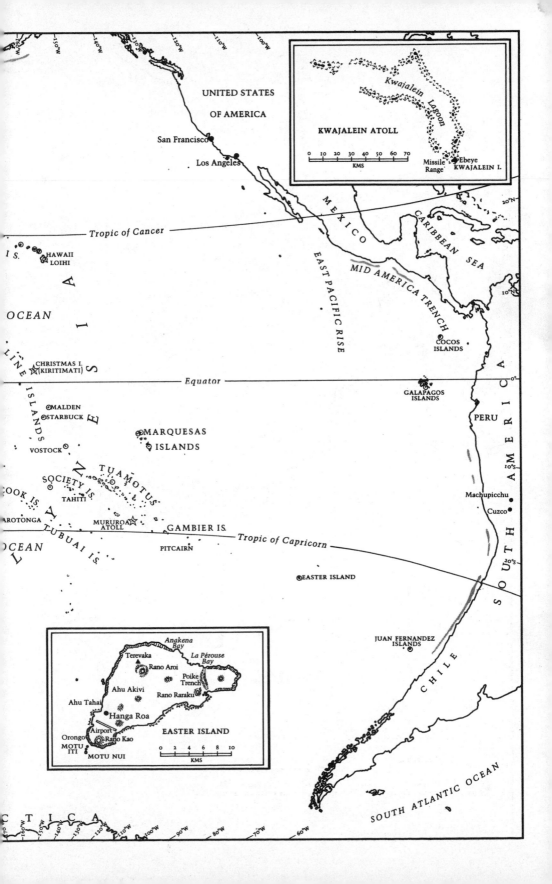

UNITED STATES
OF AMERICA

San Francisco

Los Angeles

KWAJALEIN ATOLL

Kwajalein Lagoon

0 10 20 30 40 50 60 70
KMS

Missile
Range Ebeye
KWAJALEIN I.

M E X I C O

CARIBBEAN
SEA

20°N

Tropic of Cancer

MID AMERICA TRENCH

10°N

I S.

HAWAII
LOIHI

EAST PACIFIC RISE

OCEAN

COCOS
ISLANDS

LINE ISLANDS

CHRISTMAS I.
(KIRITIMATI)

Equator

0°

GALAPAGOS
ISLANDS

MALDEN
STARBUCK

PERU

MARQUESAS
ISLANDS

VOSTOCK

10°S

TUAMOTUS

SOCIETY IS.

COOK IS

TAHITI

Machupicchu

Cuzco

AROTONGA

MURUROA
ATOLL

GAMBIER IS.

SOUTH AMERICA

Tropic of Capricorn

TUBUAI IS.

PITCAIRN

OCEAN

EASTER ISLAND

20°S

CHILE

JUAN FERNANDEZ
ISLANDS

*Anakena
Bay*

Terevaka

*La Pérouse
Bay*

Rano Aroi

Ahu Akivi

Poike
Trench

Rano Raraku

Ahu Tahai

Hanga Roa

EASTER ISLAND

Orongo

Airport

Rano Kao

MOTU
ITI

MOTU NUI

0 2 4 6 8 10
KMS

SOUTH ATLANTIC OCEAN

C T I C A

SOUTH ATLANTIC OCEAN

CONTENTS

LIST OF ILLUSTRATIONS

The First Colonizers (between pages 80 and 81)
1   Monuriki island, Fiji (F)
2   Staghorn and agaricid corals (AM)
3   Na Pali coast (AM)
4   Crane flies (F)
5   Cardinal honey-eater (F)
6   Tongan flying foxes (AM)
7   Ferns take hold on lava flows: Savai'i, Western Samoa (F)
8   Crested iguana (AM)

Island Enigmas (between pages 112 and 113)
1   Giant prehensile-tailed skink (AM)
2   Giant lobelia (AM)
3   Little honey creeper (AM)
4   Solomon giant frog (AM)
5   Collared lory (F)
6   Satin flycatcher (F)
7   Bristly liverwort grappler (SM)
8   Fruit bat (F)
9   Marsupial cuscus (TF)
10   Tree kangaroo (F)
11   Cassowary (F)
12   Blue-tailed skink (F)
13   Gecko (AM)

The Impact of Man (between pages 144 and 145)
1   Langa Langa lagoon at sunset (AM)
2   Tatooed warrior (F)
3   Statues on Easter Island (AM)
4   Beku idol at Pa Na Ghudu, Simbo Island (AM)
5   Skull shrine in the Solomon Islands (AM)
6   Navala Village, Fiji (F)

Paradise in the Balance (between pages 176 and 177)

Attribution of photographs:
(AM)   Andrew Mitchell
(F)   Frithfoto
(SM)   Steve Montgomery
(TM)   Tim Flannery

LIST OF MAPS

ACKNOWLEDGEMENTS

So many people offered their time and thoughts when I explored the islands that it is impossible to mention them all. A few stand out for their exceptional kindness in sharing ideas and providing unstinting help. To Paul and Gracie Atkins and Mike de Gury I owe special thanks for a home base and my introduction to whales. Mike McCoy's canoe trips through the Solomon Islands were dramatic, his humour unforgettable, his company splendid. Dick Watling and his wife Kelera offered copious *kava* in Fiji and assisted with many introductions, as did Paddy and Felicity Ryan, to whom I am indebted for their hospitality. I would also like to single out Isocfatu Reti, Yoshihiko Sinoto, Bengt Danielsson, Hiro Kurashina, Yves Letocart, Rick and Bonnie Steger, and the National Parks Service of Hawaii, for their help, and in particular Steve Montgomery, who managed to examine the manuscript on a field trip in the Marquesan Islands. Thanks are also due to Clifford and Dawn Frith who obtained many superb photographs, some of which appear in this book.

I should like formally to thank the following organizations and their staff for their kind help: The Smithsonian Museum of Natural History, Washington; The Bernice P. Bishop Museum; the National Parks Service; the Nature Conservancy, Hawaii; the University of the South Pacific; the National Trust for Fiji; the Fiji Museum, Fiji; the Division of Aquatic and Wildlife Resources, Guam; the Australian National University, Canberra; the Australian National Museum, Sydney; the South Pacific Regional Environment Programme, New Caledonia; EARTHWATCH, Boston; the Richard Gump Biological Research Station, Moorea; the Department of Forests, Samoa; the International Council for Bird Preservation, World Conservation Monitoring Centre, Cambridge.

I am indebted to those who shared some of their secrets about nature and man in the Pacific with me: Graham Baines, Aaron Bauer, Bob Beck, Peter Bellwood, Simon Best, Charles Birkhead, David Blockstein, Jeremy Carew-Reid, Bruce Carsen, Claudio Christino, Fergus Clunnie, Harold Cogger, Lilian Consalez, Paul Cox, Jared Diamond, Lu Eldredge, Tim Flannery, John Flenley, Raymond Fosberg, Wayne Gagne, Paul Geraghty, Judy Gradwohl, Roger Green, Vern Harris, Rod Hay, Gerry Heslinga, Jeremy Holloway, Frank Howarth, Rhys Jones, Ken Kaneshiro, Alison Kay, Art Madieros, Gerald McCormack, John McCosker, Don McNeil, John Miller, Maeva Navarro, Paddy Nunn, Robert Palmer, Brian Parkinson, Nicholas Payne, John Peak, Siki Raputuki and family, Peter Rodda, David Routledge, Julie Savidge, Birendera Singh, Victor Springer, David Stoddart, Bill Timmis, Tony Utanga, Sione Vailanu, Nigel Wace, Rick Warshauer, Ivy Watkins, Doug Yen.

Lastly, thanks to everyone at Collins, in particular Carol O'Brien and Amanda McCardie for their unstinting work on the book, and to Moira Storey and Cynthia Benitz for their tireless initial research. To the organizers of the Scientific Exploration Society's expedition, Operation Raleigh, I bear a special debt for my first taste of the South Pacific aboard the research vessel *Sir Walter Raleigh*. On this voyage I met many companions far more knowledgeable than myself and fruitful ideas were born.

A NOTE ON PRONUNCIATION

When missionaries first put the Fijian language into writing in the mid-nineteenth century it was made completely phonetic. This left five consonants available to substitute for sounds which in English are spelled with two letters. Thus:

>   *b* – is pronounced *mb* as in 'timber'
>   *d* – is pronounced *nd* as in 'sandy'
>   *g* – is pronounced *ng* as in 'singer'
>   *q* – represents *ng* as in 'finger'
>   *c* – represents *th* as in 'then'.

The great chief Cakobau of Fiji is therefore pronounced *Thakombau*, the island of Bau where he lived is *Mbau*, and Nadi where you arrive by plane in Fiji is *Nandi*. Beqa Island, where Fijians perform the ancient art of fire-walking, is pronounced *Mbenga*. Pago Pago, the capital of American Samoa, is pronounced *Pango Pango*. For the most part I have written place names in the correct local spelling.

In areas such as Tonga, Samoa and Hawaii a glottal stop is frequently used, so that in Samoan the word for a high chief is *ali'i*, and Hawaii should technically be *Hawai'i*. In New Guinea and the Solomons pidgin English is spoken. At first this is incomprehensible, but many bastardized English words soon become recognizable. For example edible greens are known as *kabis*, *aesboks* means refrigerator, and the best is *namba wan*. Most of the languages are economical with grammar and easy to pick up.

I was happy to discover that except in French Polynesia, English is spoken widely in the Pacific.

If I sail my canoe
Through the breaking waves
Let them pass under,
Let my canoe pass over,
Tane!

Tahitian prayer song

You gentle breezes of the South East, playing
   and caressing each other above my head –
   hurry!
Run together to that other island:
You will see someone who has abandoned me
   seated in the shade of his favourite tree.
Tell him you have seen me in tears.

Traditional Polynesian refrain

Chapter One **SETTING FOR A RING OF FIRE**

Ages ago, when the world was different, the South Pacific contained many islands we have never known.

James A. Michener, *Return to Paradise*, 1951

 rom the eastern shores of Japan and the Philippines to the western coast of South America, some 25,000 islands rise from the surface of the sea. Most are atolls, scattered necklaces of islets encircling a lagoon, but many are remnants of ancient high volcanoes which poke their heads above the blue surface like fluted emeralds topped with white clouds. The Pacific Ocean covers a third of the globe; as much as the Indian, Atlantic and Arctic Oceans combined. The deepest waters are to be found here, plummeting to 11 kilometres off the Philippines. Including Asia and Japan, one third of the world's population lives in the Pacific Basin. New Guinea uses some 700 languages, the Solomons 70; there are about 2,000 in the area as a whole. From east to west is a distance of 16,000 kilometres, three times the width of the United States of America.

Most of us are unfamiliar with the vast number of islands in the Pacific, imagining that Fiji, for example, is just one place rather than the diverse collection of islands that it is. There are 322 islands in the Fiji group alone. Even after the arrival of the Polynesians and the European Age of Discovery the islands of the South Pacific remained essentially isolated, but in the last decade air travel has changed all that. There are few important islands that do not possess an airstrip now. This new access to the outside world is fuelling change at an unprecedented pace, but much of the rest of the world remains unaware of it. The desire of urbanites to get away from it all is spawning a huge tourist industry from America, Japan, Australia and France. Enchanting smiles and deserted coastlines lure an ever-growing horde of would-be castaways. The first casualty will be our image of Paradise and the natural environment which fulfils it, on which many Pacific cultures depend for their well-being.

Since 1945 the US, France and Britain have been responsible for testing more than 250 nuclear devices in the Pacific. Following the detonation on Bikini of a bomb equivalent to 1,000 Hiroshima explosions, children downwind on Rongelap atoll played in radioactive dust as though it were snow. The Marshallese and others are today suffering a legacy of cancers and deformed children which stems from the British and American tests in the 1940s and 1950s. There is a continuing traffic in atomic weapons across the Pacific; America alone has over 160 military bases there. France is doing everything in her power to dissuade Pacific nations from becoming 'nuclear-free', even to the extent of blowing up a Greenpeace protest ship in New Zealand before it could meddle with French tests in Tahiti. The islanders continue to suffer the dumping of nuclear and other toxic waste in what they regard as their ocean: the producing nations consider the poison too dangerous for their own back-yards.

The desire of young Pacific nations to develop and take their place on the international stage threatens to swamp the area in dollars and yen. The most rapidly growing economies are in the Pacific. America now trades more with the Pacific than with any other region of the world. The Soviets have signed lucrative deals with Kiribati to exploit the huge marine resources within its 200-mile zone in a part of the world that until then was regarded as an 'American Lake'. American and Japanese tuna-fishing fleets use massive purse seine nets to sweep wealth from the seas. They are responsible each year for the deaths of anything up to 100,000 dolphins which accidentally drown in the nets. The huge timber resources of New Guinea and the Solomons are being chopped down and sold for export, fuelling western demand for hardwood and paper pulp. With the vanishing forests will go the wildlife of Paradise, before the world is even aware of its existence.

The isolation of islands in the Pacific has resulted in many animals and plants which are endemic, meaning that they occur in a limited area only. They are desperately vulnerable to extinction through introduced alien predators, diseases and habitat destruction in a way that more widespread species on continents are not. The Pacific islands deserve our attention – there are more endangered species here than in any other region of equivalent land area in the world. Almost half of all the endangered birds in the United States are endemic to Hawaii.

Though they arrived less than a few thousand years ago, humans have done much to change the nature of the Pacific, but the colonization of the islands began long before they ever set foot there. Millions of years earlier the islands were claimed by a diverse collection of natural castaways which evolved on the islands in conditions of unprecedented isolation. Despite three centuries of scientific investigation our knowledge of them today is very poor. The way these archipelagos were colonized in the past and the changing balance between nature and man in Paradise today make up one of the most absorbing and unexplored natural histories the world still has to offer. It is a story which begins millions of years ago with the splitting of continents and the outpouring of molten rock and fire.

Two hundred million years ago the world's landmasses were joined together in the supercontinent of Pangea which stretched from the North Pole to the South. Twenty million years later the northern half split off to form Laurasia, leaving Gondwanaland to the south with the Tethys Sea in between, the forebear of the Indian Ocean, Mediterranean, and West Pacific. Fifteen million years passed and the two sister continents were slowly torn apart from north to south, opening the Atlantic Ocean and encircling the Pacific behind them.

Far from being a solid structure, the earth's surface is fragmented like a cracked eggshell into plates which move in response to turbulence in the earth's molten core. Where the plates rub shoulders a battle of titanic forces rages; whole continents move in collision, buckling to form mountains such as the Rockies or the Himalayas. The battle is not always equal. Heavier oceanic plates lose out to those of continents and slide beneath them to melt in the depths of the earth. Where the fractured edge disappears, huge deeps are formed such as the Middle American Trench off the west coast of

Mexico or the Kuril-Izu-Mariana Trench to the east of Japan. A few miles to the east of Easter Island, surges of magma at the East Pacific Rise divide and flow east and west. The Cocos, Nasca, and Antarctic plates will eventually collide with the Americas, but the largest section, the Pacific plate on which most of the islands sit, edges at about eleven centimetres each year in a northwesterly direction. Heading for a gigantic but almost imperceptible collision, its leading edge challenges the might of Eurasia complete with the vast bulk of the Himalayas. The Pacific plate cannot win the unequal struggle and slides slowly into the Japan Trench. Further south is a second collision point, this time with the Indo-Australian plate. Here the deep trench swings east in the shape of a monkey's tail past the Solomon Islands, turning south at Tonga towards New Zealand. As millions of tons of rock pour into the earth, volcanic eruptions are fuelled on a massive scale around the plate's rim. This is known as the Ring of Fire.

The Pacific's warm tropical seas have nurtured the birth of a multitude of islands over millions of years, sandwiched for the most part between the Tropics of Cancer and Capricorn. Some, such as New Caledonia and New Zealand, are fragments discarded from ancient continents. The volcanic island chains which are scattered around the edge of the Pacific, such as the Mariana Islands and the Aleutians, are the offspring of the Ring of Fire, spewed out as recycled bedrock from the ocean floor. Those islands which rise in the centre of the Pacific plate, such as Hawaii, Samoa or the islands of French Polynesia, are formed in a different way. They are born from 'hot spots' beneath the crust which burst through to the surface like pimples on the ocean floor. Young cones build upwards through cold columns of water. After perhaps thousands of years, basalt rock, barren and lifeless, breaks the surface to touch the air amidst clouds of sulphur and steam. The largest mountains on earth, mightier than Everest, were born in the Pacific in this way. Only their jagged peaks are visible above the surface; they are known as the high islands.

Once extinct, these great mountains begin to sink under their own weight at a rate of about a centimetre a century, until eventually their eroded cones vanish beneath the sea, leaving a chain of atolls. As each cone continues to subside the coral grows upwards, forming the characteristic ring of small islands or *motu* at the surface. Sometimes the mountain might rise again, pushing the atoll out of the sea to form a flat-topped *makatea* island covered in dry coral stone with a marshy lake at its centre, the poor remnant of a once beautiful lagoon. With volcanic islands, atolls, raised *makatea* islands and the fragments of ancient continents, there were four kinds of enchanted islands in the Pacific ripe for invasion.

How animals and plants reached them remains one of the world's greatest natural mysteries. Seeds and small insects may have been wafted on air currents through the atmosphere; animals with wings may have flown. Some creatures may have reached the islands clinging to driftwood, but the distances involved are so great that it is hard to believe that they arrived by chance alone. The creatures roaming the forests and beaches today are the products of millions of years of evolution; some forms are unique to a single island or group of islands. Without competition from similar species, birds have lost their powers of flight, crabs have taken to the trees, crayfish have crawled from the sea to colonize the land, and dwarfs have turned into giants. Beneath the sea the story is the same. Isolated coral reefs have produced a unique richness, giving birth to new species beneath uninhabited atolls. Shrimps take up residence inside giant clams, or pick the teeth of passing fish, while multicoloured clown fish court safety in the poisonous arms of giant sea anemones. How could a bat, a snail, a flightless bird have possibly made such a journey? Yet these creatures are found on even the most remote islands.

Modern theorists vie for attention, torn between two opposing views. One holds that the Pacific's distant archipelagos were colonized by chance from the continents by land, sea, and air through the efforts of wind and currents, and even as hitch-hikers on sea birds: the so-called Dispersal Theory. The majority of animals and plants in the South Pacific islands, and even Hawaii, are closely related to those of South-east Asia, yet the prevailing winds and currents which might have assisted these early castaways flow in the opposite direction, from America. If dispersal was the mechanism, why should there be so few animals and plants in the Pacific from the Americas? A second view proposes that animals and plants colonized the islands across now-submerged land bridges, or simply sat still and let the land move for them: the Vicariance Theory.

A glance across a map of Oceania reveals the present position of the Pacific's islands, though they were not always where they are now. In ancient times, whole groups of animals and plants travelled great distances from their points of origin merely by sitting still, evolving on the islands and continental fragments on which they were being carried as the earth's crust moved beneath them.

To discover the origins of the Pacific's natural world, it is valuable to reconstruct the past and see how one continent was connected to another. Few plants from those ancient times survive today, but some living fossils provide vital clues. The Antarctic beech is one example. In both South Australia and the tip of South America the same species of Antarctic beech tree grows with the same insects feeding on its leaves, providing compelling evidence that these two continents were once joined by land. The piece of the jigsaw that is missing today is Antarctica. Antarctica remained connected to Australia until 45,000,000 years ago. Then it was not an ice continent, but grew with lush forests and enjoyed warm summers. Marsupials from South America journeyed through the southern beech forests to the land that was to become their evolutionary refuge. New Zealand and New Caledonia had by this time already broken away from Australia, so no naturally occurring marsupials are found there today. Following its break with Antarctica, Australia moved north with its collection of independently evolving pouched mammals and fused with New Guinea. Elsewhere in the world the marsupials were in retreat. A new kind of mammal had appeared on the earth. These mammals bore their young not in pouches, but inside their bodies. They were more aggressive and efficient and almost wiped the marsupials from the face of the earth; eventually they were to give rise to man. Possums are the only marsupials which survive in the Americas now.

Like Australia, New Guinea was protected from the onslaught of the more efficient placental mammals by the sea. Though it was once possible to walk from the Malaysian peninsula to Borneo on the Sunda shelf, no such land bridge ever existed to enable animals to walk further east. The Wallace Line, as the deep-water trench causing this division of nature has come to be known, has forever remained a turbulent water barrier to commerce in creatures between the two regions. The woolly rhinoceros was condemned to roam the forests of Borneo and not those of New Guinea, and beautiful clouded leopards had to remain content with pigeons, being unable to reach the tasty kangaroos that hopped about Australia's woodlands. No large mammals are to be found in New Guinea's forests now, though once there was a giant, almost panda-like marsupial bear. It may be that man the hunter accounted for them all when he reached the island some 45,000 years ago. The sea, however, was no barrier to plants. Some seem to have their ancestors as far north as China. Seeds travelled well over water in the guts of birds. On New Guinea's mountain slopes, just south of the equator, northern oaks abound, dropping acorns into the tropical soil.

This mysterious island lies at the crossroads of two of the great zoogeographic regions of the world, the Oriental and the Australian regions. About 22,000 species of plant exist here, of which 90 per cent are found nowhere else in the world. More than a quarter of the flowering plants are orchids, so the forest branches are festooned in season with beautiful flowers. There are 252 different reptiles and amphibians in the mountains, including spectacular tree frogs, marine and freshwater turtles and monstrous saltwater crocodiles. Some of these last reach a length of over eight metres and have been known to attack canoes several miles out to sea. In addition New Guinea has numerous birds; the birds of paradise with their plumages of metallic blues, bronzes and greens are some of the most beautiful in the world. Most of the birds are related to those of Australia, though some, such as the Papuan hornbill *Rhyticeros plicatus*, sandpipers and tree swifts, came from the Oriental region. Off the south coast, at Port Moresby, the greatest diversity of corals in the world is to be found. This huge island and its even larger neighbour to the south, to which it was recently connected by dry land, seem to have been the evolutionary melting pot from which many of the Pacific Islands were colonized. The journey to the east was a blend of evolutionary challenge and chance in which some would succeed while others would be for ever confined to their Papuan refuge. The great distances across the Pacific acted as a filter which only a few animals and plants have been able to pass. Those plants that did found islands ripe for exploitation, virgin volcanic soils with minerals for the taking, and later provided fruits and berries for birds to seize without competition from other hungry bills. From Australia, then one with New Guinea, came colourful pigeons and songbirds. Insects and lizards moved with more ancient continental fragments of Gond-wanaland such as New Caledonia, and a selection of creatures and plants began to disperse to the Pacific's emerging volca-noes. Evolution here moved in leaps and bounds, breeding variety from the few initial colonizers.

The tropical Pacific is also home to three main groups of people: the Micronesians, who inhabit the vast numbers of atolls and islands to the north of New Guinea; the Mela-nesians, who broadly range from New Guinea to Fiji; and the Polynesians, who colonized a great triangle of ocean with Hawaii at its apex and New Zealand and Easter Island at the bottom corners. Ever since Captain Cook first sailed the Pacific, the origins of these people have fuelled even greater debate. To early explorers it seemed impossible that Stone-Age cannibals, as many of them were, could possibly have reached these remote archipelagos in dugout canoes without sophisti-cated navigational equipment. Divine intervention was prop-osed, and even vanishing continents.

A nineteenth-century French physician called Augustus Le Plongeon invented a lost continent in the Pacific which he called Mu, believing that he had evidence to show that the Muvians were the ancestors of both the Maya and the Egyptians. An Anglo-American, James Churchward, believed Mu to be an extension into the Pacific of a sunken continent in the Indian Ocean named Lemuria. He took the flooded city of Nan Madol in the western Pacific and the extraordinary giant statues of Easter Island in the east as evidence for this theory. Churchward believed that Mu disappeared mysteri-ously beneath the ocean about 13,000 years ago, leaving only the myriad islands of Polynesia to indicate that it was ever there.

Madame Blavatsky, a nineteenth-century occultist whose hobbies ranged from circus bareback riding to obtaining mystic wisdom from the feet of Tibetan mahatmas, thought a new Atlantis had been found. She believed that the Lemurians were one of the routes for the evolution of mankind: giant ape-like creatures, some with four arms and third eyes at the backs of their heads. Descendants, according to the occultists, survive today – as Hottentots, Aborigines and Papuans. Later descriptions were even more bizarre. William Scott-Elliot announced that the Lemurians were four and a half metres tall, with flat faces and eyes so far apart that they could see sideways like birds, and protruding heels which enabled them to proceed forwards or backwards with equal ease. Originally they were egg-laying hermaphrodites.

Since then our knowledge of the origins of Pacific peoples has increased greatly, but mysteries remain. Powerful seafaring civilizations developed in a number of places. One of the most extensive was that of the Yappese Empire in Micronesia which stretched from Palau, close to the Philippines, across the Caroline Islands towards Pohnpei, formerly Ponape. Here hundreds of giant coins made of glittering stone lie propped against steep-roofed meeting houses. The largest stones require ten men or more to lift them and there is no source for such rock on Yap. In the Palauan Islands, 500 kilometres away to the east, unfinished stones cut from stalagmites have been found in caves. Somehow the massive stones must have been transported by raft or canoe to Yap where they served as Chiefly symbols of power. Of the customs and origins of these people almost nothing is known. The ancient city of Nan Madol on Pohnpei is described as the Venice of the Pacific. Its water-filled canals wind through eighty artificial islands flanked by impressive basalt walls and terraces built in the thirteenth and fourteenth centuries. Though the people who built it may have had their ancestors in the Philippines, and it was the major centre of a Micronesian Empire, there is still evidence today of the effects of its power.

Moving still further east another mystery lies on the island of Tongatapu. Here a massive limestone trilithon arch stands alone like a single gateway to a Polynesian Stonehenge. Legend has it that the god Maui carried it to the Tongan kingdom from the Wallis Islands, but it seems more likely that the 11th Tu'i Tonga erected the 110-tonne monolith in about AD 1200.

Perhaps it was a giant entrance to the now vanished royal compound. The present King of Tonga proved that markings on the lintel could be used for predicting the equinox when observed at sunrise on the summer solstice; this may therefore have been its purpose.

Thor Heyerdahl's pioneering work in the 1950s strongly suggested that the Polynesians were American Indians; others oppose that view, believing the origins of all Pacific peoples to have been in South-east Asia. To me these intriguing ideas, coupled with the popular image of the South Seas as a Garden of Eden, presented an irresistible package. My aim in this book has been to follow in the footsteps of the natural world and man by journeying from west to east across the tropical Pacific, and to explore four themes. How did animals and plants reach the Pacific's distant archipelagos? How did these castaways evolve to give us the unique collection found on the islands today? What happened to these island environments when humans first arrived? What state are they in today?

Nowhere illustrates the plight of Pacific islanders in a modern world better than a curious speck of land in Micronesia in the western half of the Pacific: the island of Nauru. There would be no reason to visit Nauru but for two facts: its airline provides access (albeit unreliable) to otherwise inaccessible parts of the Pacific, and its inhabitants are some of the wealthiest people on earth.

Nauru is a small lump of coral limestone raised less than 70 metres out of the sea, lying just south of the equator to the north of the Solomon Islands. It is oppressively hot for much of the year, and isolated on all sides by vast tracts of ocean; the nearest land is the equally uninspiring island of Banaba 320 kilometres to the east. The Solomon Islands lie 1,500 kilometres away to the west. Normally it would be extremely difficult to reach. Scrubby green vegetation covers much of the interior, and modern bungalows that have seen better days litter the thin coastal strip. A mere hundred metres wide, this is the only place where cultivable soil exists. The people who live here are mostly very fat, move with great economy, and have a reputation for an indifference to other people which borders on unfriendliness.

The source of the island's wealth lies in the fact that for centuries vast numbers of seabirds roosted here. The resulting piles of bird excrement – fortunately not fresh, but thousands of years old – have turned to a form of soft rock known as guano. This built up to huge depths before the birds abandoned the island, perhaps when the first humans arrived. The first Nauruans were unaware of the value of what they were sitting on until an Australian named Albert Ellis noticed a peculiar rock being used as a doorstop at his trading company offices in Sydney. It had been sent there in about 1888 from Banaba with a suggestion that it might be good for making into marbles, and turned out to be almost pure phosphate, a product of enormous value as fertilizer. Britain quickly annexed Banaba and the British Phosphate Commission (BPC) signed a paper deal for mining rights at a mere £50 a year for 999 years. A series of paper agreements gave the Banabans no significant rights whatever, and finally, after World War II, they were forced to abandon their ruined island and resettle on Rabi Island in Fiji. It was not until 1981 that the Banabans received $A 10,000,000 in compensation from the British Government. The Nauruans were sitting on a similar goldmine, and for half a century were similarly exploited by the BPC, but in 1968 they became independent and struck a better deal.

Flying into Nauru is a terrifying experience. Air Nauru, the islanders' own airline of three Boeing 737s (there used to be six) is a wonder of the modern world. In splendid blue and white livery the airline operates a heavily subsidized network which provides the only quick access to large areas of Micronesia and the West Pacific. The only problem is that having a ticket is no guarantee of a seat. Furthermore, there is no easy way of knowing when the planes will fly. The Chief of Nauru has been known to disrupt the schedule by sending a plane to Sydney to collect frozen prawns for a party. A plane full of enormous Nauruans is incapable of getting off the ground, so their numbers are monitored carefully and numbers of available seats reduced accordingly. Pilots, mainly Australians, are forced to beg safety manuals from other airlines, and have to load the planes themselves whenever the Nauruan staff finds the tedious process all too much. The airline does have an excellent safety record. Nonetheless, flying at night into the short palm-fringed airstrip on my way back from Guam, my heart was in my mouth. I was marooned on the island for a couple of days to wait for an onward plane, so I booked into the Meneng Hotel, the only one of any size on the island.

Guests appear to be an inconvenience at the Meneng. At the reception desk no one was to be seen. Eventually a lady of considerable size reluctantly offered me a room and called to an enormous porter asleep on his back on a bench to carry my bags to my room. He did not stir. The Meneng is built on the coast in the face of the wind which pours over the reefs; the pounding surf contributes a mist of salt spray which is slowly corroding the hotel. Door handles are flaky green with decay and hinges are rusted almost solid. The 'Speedie' kettle took thirty-five minutes to boil. Cockroaches enjoyed races over the deep-pile carpet. The maid who replenished the towels wore a Rolex watch.

The restaurant at first looked promising. 'Ah, the oysters are off. What about *boeuf florentine*?' The waitress examined the menu closely as if amazed that anything was on it at all. She announced that she had never heard of the *boeuf florentine*, so it too must be off. The dover sole? More confusion. I played safe and ordered 'local fish' and a bottle of Chablis. Claret arrived, but it seemed churlish to complain. The whole thing cost just £2.50. A Japanese businessman looked perplexed as a waitress in bright-orange uniform shouted 'You've had enough – that's all you're getting. You've had too much already.' This was her response to his request for a third cup of tea.

The following day I decided to explore the island. In the only small town shops bulged with consumer goods and piles of tinned fish. A few houses sported solar panels. Some Nauruans crowded around a radio in a betting shop, placing money on horses in Sydney. The local draw was worth $A 2,500 that day; some days it is as much as $A 100,000. Many play the stock market, but broken air-conditioners hang from house walls. The houses are provided by the Government, which also covers most household bills such as those for the telephone and electricity. The royalties from the phosphate mines provide Nauruans with the highest per-capita incomes in the world, higher even than those of the Saudis. The wealth here is spread around the mere 5,000 people who live on the island. Shining motorbikes purr along the single main street, pursued by extremely ancient and spluttering open Land Rovers fitted with gigantic stereos.

Huge cargo ships bring water at $A10 a tonne from Australia and return full of dried phosphate delivered from pipes held by two huge gantries which march across the reefs from the processing factory on the coast. The Nauruans draft in Kiribati people to do the dirty work in the mines at $A2 a day.

As I ambled up the hill in the sweltering heat towards 'Topside', as the area where the mines are worked is called, a green Land Rover clattered up beside me and a Nauruan peered out, a pair of Walkman headphones straining across his glistening temples. 'Would you like a lift?' He beamed with unusual friendliness.

I swung in and we rumbled off along the dirt track into the interior. It was a tour of desolation. Huge diggers on caterpillar tracks clawed at the rock between the coral outcrops that lay buried beneath the guano. Enormous lorries roared away with topsoil and trees to a nearby dump. Where the mechanical diggers had passed, all the original forest had been peeled back and bare limestone pinnacles the height of four men pointed to the sky. Where the guano had been removed from the spaces between them, a jagged and spectacular moonscape remained. A solitary noddy tern flew slowly over the extraordinary scene as if trying to find a place to land. I could imagine the place as it once was, teeming with the sound of a million sea birds. To rehabilitate them on the island would cost millions of dollars, even if it should be possible once the mining has ceased and the scrub vegetation has taken over. The mining began in 1905; a minute royalty was paid to the Nauruans by the British Phosphate Commissioners until 1965. Nauru became the world's smallest independent nation in 1968 and has run the mining operation itself since 1970. Since that time the 2,000,000 tonnes or so it exports each year has made it rich, but there is only enough phosphate for about another nine years.

'What will you do when the phosphate is finished?' I asked my companion who worked as a driver for the mine.

'Go back to the old ways.' He gave a nervous grin.

This will be hard for most Nauruans, who have grown accustomed to an easy life. Few now turn the coastal soils for taro crops or coconuts; eighteen holes at the local golf course seems a better way to spend one's time. Time is the one thing most Nauruans have on their hands. At two places on the island they spend it on an ancient game whose origins must go back to earliest Polynesian days. There are few native birds on Nauru, but sea birds still patrol the coasts, and the Nauruans like to try to catch the man o' war or frigate bird for sport. These enormous birds soar on great black wings, streaming forked tails behind them. They are parasites, living off fish from the stomachs of other sea birds returning home, which are forced to regurgitate their meals in spectacular feats of flying.

As they pass the time of day over cards and beer, Nauruans keep a lookout for approaching frigate birds. If they spot one, tame, juvenile frigate birds are stirred into the air from a special wooden frame nearby, on which they have been trained to sit, and begin diving for fish the islanders throw into the reef shallows for them. This activity attracts the adult birds, at which the Nauruans sling lead weights attached to nylon fishing-lines as they approach, hoping to entangle their wings and pull them to the ground. Fortunately, they are rarely successful and as often as not return to their card-playing amid much guffawing and laughter. After all, another bird will come by tomorrow.

Evening was approaching as I walked back towards the small township after watching one such attempt. Women played basketball near the coast, shrieking with laughter as the sun spread sheets of gold across the calm inside the reef. A rhythmic boom came from the dilapidated Baminda Inn. I peered through the broken windows; inside was an empty dark room reeking of urine and scattered with a few broken chairs. In a dark corner five young drunks huddled on benches next to a huge ghetto-blaster. On seeing me they waved unsteadily and bawled for me to join them. I hurried on. It was one of the saddest sights I saw in the Pacific.

Nauru airport is unlike anywhere else I have been to. Departure was scheduled for 11.00 p.m. – which meant very little, of course. In a small building under the palms, an adolescent in T-shirt and shorts examined the tickets. This is the hub of Air

Nauru's network in the Pacific. Indians greeted each other with nodding heads; Sikhs were talking politics; Kiribati women with children were laughing and smiling; Australians were checking their watches; dark Solomon Islanders sat and stared, together with French servicemen on their way to New Caledonia. Several children were playing tag around the ticket desk. Their elephantine mother, in a voluminous white dress printed with huge blue hibiscus flowers, occasionally wobbled over and hissed at them in a way which was apparently designed to make them stop. With renewed vigour they clambered on to the luggage conveyor and disappeared through a hole in the wall, no doubt on their way to the aircraft's hold.

Three hours later we lifted into the night, leaving this extraordinary nation behind. It seemed to me that development had provided everything the islanders could have wanted and left them with nothing at all. Argument now rages over who should pay for the rehabilitation of their ruined home. The plan involves shipping millions of tonnes of earth there and dumping it in the holes where 100,000,000 tonnes of bird droppings used to be. It will be easier to buy a new island. The Nauruans have considerable investments abroad, including the tallest building in Melbourne (affectionately known locally as Birdshit Tower), but the collapse of phosphate prices and the Australian dollar has left them wondering about their future. I settled into the comfortable blue and yellow seats of the aircraft, each fit for one Nauruan buttock, and wondered too.

As we approach the 1990s, all is not well with this oceanic Eden; much of it is ignored and growing with weeds. Thankfully it still contains the most beautiful islands in the world, and most authors will continue to praise the good looks of the islanders, the sweet smell of the flowers, the brilliance of the glittering coral reefs. But within the islands' forested and little visited interiors there are equally beautiful inhabitants whose story has never been told and whose lives are now more at risk than ever before. This book is a tribute to these forgotten islanders, on the brink of a new Pacific age.

Chapter Two

BRIDGES TO THE PACIFIC

The Solomon Islands

> Heaven no longer monopolises our concerns.
> Attention is turned toward the earth, which
> has to be explored, toward man, who has to be
> known.
>
> Hubert Deschamps, 1749

hrough gaps in white spray, the island of the head-hunters beckoned with an irresistible menace. Almost hidden from the world behind mists that rose from the surface of the sea, Simbo lay on the horizon like a crouching turtle, a grey outline against the stormy Solomon sky. It is a small volcano some thirty kilometres from Gizo, the island capital of the Western Province. My knuckles were drained of blood from the constant strain of gripping the sides of the canoe as we sped through the increasingly rough waters which tossed the surface of the Pacific swells. In front of us a mountain of white water was building on the crest of one huge swell, and as we climbed towards it, the summit curled, roared with a wicked pleasure and rushed towards us, licking its turbulent green surface with hungry tendrils of foam. As I peered ahead I began to feel certain that we were soon to drown. With a huge crash, island and sky were obliterated in a cascade of white spray that filled the air around us and bit into unprotected eyes already reddened and smarting. The outboard whined in protest as the canoe flew into the air through the top of the wave; briefly I glimpsed the sea all round tossed with white crests, then with that sickening roller-coaster feeling we began to drop into the deep trough on the other side.

It is at moments like this that I wonder why I am not at home, comfortably watching television. McCoy was to blame, of course. Earlier I had seen him grinning, even leering, from beneath his oilskin; with dark sinister features, full lips, watchful eyes and a glint of gold at his ear, he was a convincing picture of a true South Sea Island villain. Simbo, he said, was a short trip across the ocean by canoe; no problem – the locals always do it. Few things were a problem here, least of all being lost at sea; it was part of life and had always been so. It was not unusual for islanders to drift with broken outboards or torn sails for weeks until they made landfall on another island downwind, sometimes alive, sometimes starving, sometimes blistered by the drying sun, occasionally dead. I tried to force myself to think of something other than being lost, drowned or eaten. Solomon Island waters are notorious for sharks.

We hit the base of the trough with such force that the deck-housing across the bow cracked. The vertebrae in my spine wriggled like the links of a bicycle chain and sent riveting pain signals up into my skull. With a lurch the canoe tipped sideways, momentarily unbalanced, and green water poured over the side. It was so warm, almost welcoming. Then the propeller gripped the sea, and we surged forward again, climbing to meet yet another mountain of green that had travelled a thousand miles across the Coral Sea, sculpted each day by the south-east trade winds.

Nathan, the fourteen-year-old fair-haired son of a missionary, equipped with baseball hat, T-shirt and shorts, fought a constant battle to guide us safely, if not comfortably, towards our destination. He had been plying the waters between the Solomon Islands for most of his life and was as skilled as any of the natives. Accompanying me to Simbo Island were Patrick Purcell, an American volunteer worker with the Government Information Department, laid-back, lanky and bearded, and Mike McCoy. I had met Mike for the first time at Henderson Airport outside Honiara, the capital of the Solomon Islands on Guadalcanal.

'Pleased to meet you,' he had said, thrusting a hand towards me. 'Got your letter. Thought I'd meet you here.' I looked at the shell necklace dangling over his chest. Hair sprouted thickly from the openings in his green and black tropical shirt.

His *bête noire* was a small red four-wheel drive Suzuki jeep, which had the distinction, I came to discover, of only going up steep hills in reverse gear. Gravity would then keep the petrol

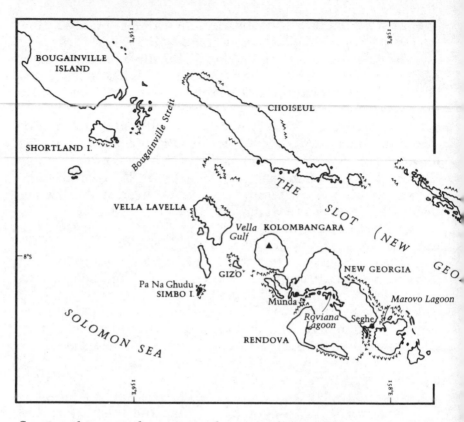

flowing down to the engine, the pump having long since given up. The harsh sun had faded the colour to a lurid pink, and most of the outside panels enjoyed the most tenuous of links to the interior chassis. Mike McCoy is the only person I have ever met in the tropics who moves with the frantic speed of someone in the London rush hour. The small congested tropical roads in Honiara were to him as frustrating as any overloaded six-lane motorway.

'Bastards!' He stood on the horn which emitted a strangled beep as we swung into the traffic, narrowly missing several passers-by with piles of vegetables on their heads. With a series of savage gear changes, and a screaming 500 cc engine we rushed down the rough asphalt road past lines of palms and trees covered in red crabs-claw flowers along the waterfront towards the town, in a clatter of loose panels and jiggling bumpers.

The steep flanks of the mountainous interior of Guadalcanal Island rose to our left. The dark forest which clothed the hills was dotted with light-green patches, which were farms. Many Americans had died there on Bloody Ridge, holding out against a Japanese army determined to remove them and advance into the Pacific. In early August 1942 the Americans captured Henderson Field, built by the Japanese as their springboard to Fiji and Samoa. Over the 200 days of the Guadalcanal campaign, 5,600 Americans were killed or wounded, and of 36,000

Japanese, 15,000 were killed in action; malaria and battle fatigue claimed 9,000 more. It was some of the most ferocious and desperate fighting in history. Amongst the banana palms and taro patches, grenades, old boots, helmets and guns still lie where the men carrying them fell.

Those terrible scenes are gone from Honiara today. On my right as we bumped into town, numerous small coastal launches and canoes laden with nets plied the calm waters of the harbour front. Islanders in brightly patterned cotton dresses which contrasted with their black skins travelled the side-walks on bicycles or bare feet, carrying bundles of wood, thatch, fruit or poles laden with coconuts for the market. A profusion of modern bungalows and white shops with flat roofs stretching over the pavements to provide shade loomed ahead. In the distance I could see other islands across the sea.

Mike and I relaxed for a while in the comfortable, traditional surroundings of the Mendana Hotel. Garishly-coloured tourists were draped across the furniture; dark, fuzzy-haired Solomon Islanders shimmered between them carrying silver trays of drinks. Across the room, through the open walls, between the palm roof and strong wooden pillars, the tropical blue sea glinted in the sunlight.

Mike's house was a prefabricated grey-white box on stilts with a flat roof, perched on a hillside like much of the cheaper housing in Honiara. Large trees overhung it, providing shade and an air of privacy. As we climbed up to the house a dog bounded down the path to meet us, and despite Mike's abusive words of greeting jumped up at us with undiminished enthusiasm. Four small wooden steps led up through the mosquito-screen door, and into the single main room, off which there was a study filled with books, where the children slept in bunks, a bedroom, and a shower room.

'You're on there,' he said, pointing to a cane sofa draped with intricately designed blankets. A strange-looking fish stared at me from a bubbling aquarium next to the window by the door. I stretched my arms upwards, sighed and narrowly missed amputating a hand on the rapidly turning fan hanging from the ceiling.

Mike had found a perfect niche in the Solomon Islands. His

delightful wife Annie was from Malaita, one of the largest islands in the chain of six that includes Choiseul, New Georgia, Santa Isabel, San Cristobal and Guadalcanal, and stretches to the south-east of New Guinea. Mike was an expert on reptiles and had studied the distribution of lizards and snakes in the Solomons and elsewhere in the Pacific, as a lesson in dispersal and evolution in the islands. A superb professional photographer, he travelled widely around the Pacific on assignments for the *National Geographic* and others, to capture the delights of life above and below the sea. I could not have asked for a more knowledgeable guide or a more entertaining companion.

With a huge crash I was jolted back to reality: the canoe plunged into another wave, sending a deluge of water over my thin green anorak, filling the bottom of the boat with water. Nathan grinned from the stern, his glasses covered in droplets, as he perched like a music-hall monkey on the rear thwart of the fibreglass canoe. The sea appeared to become a little calmer now; clouds of black terns and white-capped noddies, accompanied by an occasional majestic brown booby, skimmed over the water, searching for small fish and shrimps. Occasionally the water surface boiled as a shoal of tuna lunged at unseen prey. 'Albatrosses,' said Mike, gesturing at the birds with his beer can. All birds were albatrosses to him.

The Spanish explorer Alvara de Mendaña was the first European to arrive at Santa Isabel in the Solomons, in 1568. Inca legends of islands rich in gold 600 leagues west of Peru had drawn him there, but a fleet of welcoming war canoes presented him only with cooked pieces of a small boy garnished with taro leaves. He later arrived at Guadalcanal and found gold in the rivers of the interior. It is still mined today. The find prompted him to name the islands after the legendary African King. Defining longitude was a major difficulty for explorers of the day, and following this discovery the Solomons were lost for a further two centuries before anyone realized that Mendaña had placed them anything up to 700 leagues off position, depending upon which map was used. In 1768 Dalrymple confused them with New Guinea, and in the same year Bougainville gave his name to an island further north which was already called Choiseul, not realizing that it was part of the Solomons. Surville and Shortland were similarly mistaken when they visited the area in 1769 and 1788 respectively.

Soaking wet and bruised from the choppy swells we rounded the northern tip of Simbo into a small, shallow bay, at last in the calm. The island measures about twelve square kilometres and is the westernmost island in the New Georgia group. It was originally called Eddystone by Shortland, but now only a small pinnacle off the southern tip bears that name.

In 1894, following exploitation by European slaving expeditions, the 'Blackbirders', from plantations on Fiji and Queensland, the Solomons became a British Protectorate and they only gained full independence in 1978. They cover an area of 27,556 square kilometres, and include, apart from the six large islands, twenty other medium-sized ones and numerous islets – there are over 900 in all, stretching 1,000 miles from Bougainville Strait to Tikopia. Most of the people are Melanesians, but with soft elegant faces, rarely displaying the heavy hook-nosed features of the Papuans.

Solomon Islanders must have the darkest skins in the world; in some cases they are almost an indigo blue. People are blackest in the Western Province, with the numbers of tan-skinned islanders increasing to the east. Their dark skins are made doubly striking by the fact that many of them have bushy blond hair. Unlike the wavy black hair of Polynesians, that of Melanesians is frizzy. Where the genes originated that turn Solomon Islanders' hair any shade from black through reddish brown to the colour of bleached straw is something of a mystery. Visits by early Europeans is a hot contender but the same effect has not occurred elsewhere in such profusion, and it is likely that blond Solomon Islanders were in evidence long before their first contact with Europeans. Many modern visitors imagine that artificial bleach is the cause, but the colour of the pale hairs on arms and young faces that give these people a golden quality in the evening sun is real. I had also seen blond-haired Papuans on the northern coast of New Guinea. Like the incidence of lighter skin, that of lighter hair in the Solomon Islanders increases to the east and is most common among the Malaitans. Today the population of the islands numbers about 280,000; half of it is under the age of fifteen.

Some of the outlying islands such as Rennell, Bellona, Ontong Java and Tikopia – the so-called Polynesian Outliers – were colonized by Polynesians. It is believed that these people came to the islands from Wallis and Futuna to the east of the Solomons some 1,500 years ago. In the 1950s thousands of Micronesians, principally from the Gilbert Islands to the north-east, now part of Kiribati, were resettled in the suburbs of Honiara and Gizo. A number of Europeans and Chinese traders scattered across the archipelago complete the picture.

The Solomon Islands represent one of the increasingly rare places in the world that have not yet been 'discovered' by the tourist trade. Despite the obvious financial rewards of hordes of overweight Americans, windsurfing Australians and camera-clicking Japanese, the Government pursues a cautious course with the harbingers of lotion-lubricated change to their beaches. The number of large hotels and purpose-built 'cast-away' islands is gratifyingly small, though pressure for the floodgates to open is building.

To me the islands were the most beautiful I visited in the whole of the South Pacific. High mountain ranges rise from enormous fringing lagoons of deepest turquoise and aquamarine, into which are set hundreds of tiny, circular, coral islets, each topped with green forest and encircled by a thin strip of white sand, breathtaking from the air. The majority of the Solomon Islanders live on a subsistence agriculture centred on small rural villages often on the coasts; they can also gather fish by long-lining or netting along the coral reefs. Yam, taro, sweet potato and cassava provide the daily bread, grown in individually owned 'gardens' cleared from the forest. Despite centuries of cultivation rain forest still covers much of the land.

Between patches of open grassland marking ancient gardens, trees grow up the slopes all the way to the top of Simbo, their trunks visible through the green foliage like the pale legs of sheep. We approached a small colony of twenty or thirty huts the colour of straw yellowed in a summer's sun, newly thatched with sago-palm leaves. This was the village of Tapurai, which was originally built as a temporary settlement in the northern tip of the island but had flourished under missionary influence. As we came near the shore, the clarity of the water and the variety of the corals were astonishing. I looked down as though hanging in space. We negotiated an awkward cut in the reef, and poled our way between huge reddish-brown table-top corals, brain corals with their undulating patterns, other crusted balls of green and magenta, acroporas growing between staghorns. Small fish painted a brilliant yellow flashed through, while others streaked with electric blue paused in small groups on table-top corals to watch us glide by. Already a horde of eagerly shouting and smiling dark children were running down the beach, pale soles sending up sprays of white sand, blond and black hair bobbing, arms and elbows waving. The prow of the canoe gently crunched into the sand, and the sound of the children's whispering filled the air.

Simbo is not too far from the Pacific's ring of fire; it is still an active volcano. To the north are the rolling hills and sediments of former volcanic activity that make up Patukio Hill. The southern half of the island consists of two young volcanic cones about 400 metres high. The southernmost cone of Matindini is split from north to south into Ove on the western side, which is thermally active, and Karivara to the east.

Some older individuals appeared from the village. My reason for wanting to make what had turned out to be a rather terrifying trip to Simbo was not only to see what remained of the ancient cult of the head-hunters for which the island had been famous, but also to learn of a more peaceful pastime that the islanders now pursued, the farming of the extraordinary megapodes or incubator birds. These have evolved a unique way of avoiding the tedious and potentially dangerous business of incubating their eggs: they let the volcano do it for them by laying in the centrally heated soils around the volcano's steaming vents.

The villagers said they knew of the place but we would have to call at the next village to get permission from the Chief's son. Ahazi Keti, one of the older men of the village, said he would accompany us, and we set off again. Further down the coast we entered an impressive bay dominated by steep slopes rising hundreds of metres from the sea. In a saddle between them, the village of Legana lay. There was a small concrete jetty sticking out from the shore to which we moored. Colourful villagers emerged from huts a little way up the shore and wandered on to the jetty. There was a moment's conversation, then everyone began yelling at the mountainside to our right. In the distance up the slope a man could be seen running through the newly cut gardens.

Some minutes passed before Siso, a tall lanky islander, arrived. He was the Chief's son. A long debate ensued, with much gesticulating between Mike, Siso and others on the jetty and occasional comments shouted from a group of women on the shore. No, he was sorry, it would not be possible for us to visit the megapode site; there was a dispute over land rights and our visit would create awkwardness. My heart sank. It seemed that one family owned the land on which the birds laid their eggs and someone else owned the land over which the path went to get there. These two parties were not speaking to each other, the owners of the path believing that they should receive payment from anyone crossing their land to obtain the birds' eggs. The other group disagreed, claiming they had sole rights. An annual visit by a tourist ship had recently brought matters to a head: everyone wanted a cut of the profits. The whole business was causing Siso headaches.

'The women behind me, they own the place,' Siso said apologetically. Long pauses and glum faces on our part appeared to be having little effect. Suddenly, there was a change of heart, though Mike could not explain why: 'You can go if you pay twenty dollars.' We did and amidst smiles and waves put out into the bay again to continue down the coast.

We landed at a place called Ove Lavata, dragged the canoe over the white coral fragments which covered the beach, and headed into the forest. In many parts of the Pacific, islanders believe the spirits of the dead depart for another island. Keti told me that Ove was believed to be one of the places to which they came. Travelling parallel to the coast a short way inland we came to a small lake, surrounded on all sides by thick vegetation. Steam appeared to be rising from its clear emerald-coloured surface. I dipped my hand in: it was hot. Further on the smell of sulphur began to fill the air; the volcano's active breath was seeping unseen through the forest. Then quite unexpectedly we saw them: two megapode birds picking their way across the path in a leisurely fashion, like a pair of small black guinea fowl with over-large orange feet. The dark-brown feathers on their backs blended with the forest floor. Their heads were grey, slightly crested and equipped with bright yellow beaks. The dry leaves rustled as they disappeared into the undergrowth, perhaps having recently visited their communal nesting grounds. I was elated; to have observed these somewhat elusive birds so easily in the wild was an unexpected bonus. A few paces further on I saw one of the megapode's greatest enemies prowling across the forest floor, its sharp claws puncturing the leaves. Its body, as long as my leg, was streaked with black and decorated with spots of yellow. Its forked tongue collected molecules from the air and tasted them on the roof of its mouth as it searched for the megapode nests. It was a monitor lizard, hoping to dig into the sand and find the large protein-filled eggs. This would not be easy, for the eggs are laid at considerable depths. This explains the birds' large feet, which are used to excavate the burrows.

With so much turbulent sea surrounding the islands here, I wondered how these creatures had managed to reach Simbo. From New Guinea to the Solomons there is an almost continuous chain of islands stretching through New Britain to Santa Cruz. These form a fragmented land bridge to the Pacific down which many of the animals and plants that colonized the islands further to the east must have passed. But only some were able to cross. Although many creatures once roamed the plains that connected the great islands of the Sunda Shelf and joined Sumatra, Borneo and Java to the Malaysian Peninsula, the sea prevented large mammals crossing the Wallace Line to New Guinea and Australia. Most of the marsupials and other creatures which evolved there were likewise prevented from easily reaching the Solomons by water gaps.

New Guinea boasts 780 birds, but the Solomons muster only 150, ranging in size from tiny honey-eaters to large birds of prey such as Sanford's eagle. Colourful parrots and pigeons fly through the forests of the Solomons, as well as numerous insect-eating bats and fruit-eating flying foxes, and some of the 70 species of reptile that inhabit the islands, from turtles to lizards, are also to be found. On land there are but a few mammals: a possum, a tree kangaroo, and a collection of extraordinary giant rats. These may grow to the size of rabbits and have equally soft fur, but they are now very rare; one species has remained elusive since it was collected almost a century ago. Off the coast one occasionally sees whales, and more rarely the lugubrious features of the dugong, or sea cow. These have been heavily hunted in the past both for their meat and for their valuable tusks, whose ivory can be carved into ornaments. The habit these walrus-like beasts have of bobbing up and down in the water displaying somewhat human-looking breasts has given rise to the suggestion that they may have been responsible for the mermaid myth, but anyone who has taken a close look at the facial features of a dugong will quickly realize that only truly desperate men could imagine such a thing.

I could hear voices ahead, and we soon came to a clearing in the forest. All around, the earth was honeycombed with huge burrows, and at one end there was a small shelter made of palm leaves from the screw-pine, also known as pandanus. From one of the burrows protruded a human bottom and a pair of legs between which showers of black earth flew at intervals. The owner emerged from the burrow, his face covered in soil, and gave us a broad grin. In his hand he had a large brown egg. As many as a hundred eggs may be collected each day from the communal nesting area in this way. Surprisingly the population of the birds has remained stable here, and even increased. In other areas megapode birds are threatened as demand for the eggs outstrips the supply. Traditionally, local customs provide checks and balances which serve to protect many of the resources a village needs. Too often I was to find on my later travels to the islands that these have succumbed to modern influences and no longer apply.

Megapodes do not make caring parents. Giving up incubation to a volcano or a pile of slowly decomposing vegetation releases the parent birds from the dangerous business of sitting on their eggs where a predator may snatch them. But there is a price. The new chicks have no adults to protect them or teach them how to find food. They must therefore be advanced enough to fend for themselves and programmed to begin life on their own as soon as they hatch. Under natural conditions, they hatch fully feathered, burrow to the surface, and immediately run or take flight into the forest. This remarkable degree of development in the egg is costly. The eggs are therefore large and filled with protein in a rich yolk that takes up practically the whole interior of the egg. This makes them a valuable food, so the locals farm them. To encourage the birds to return to the same site, the sand and soil is prepared and kept soft and easy to dig. The small shelter I had seen was built to protect a favoured site from the rain, keeping the soil dry and workable for the birds. The volcano kept the air inside the burrows hot and humid, forming perfect incubation chambers.

The nesting site was close to the lake edge. Several canoes had been drawn up; easy transport for the journey with the day's crop of eggs back to the beach landing spot where they would be hauled over the bank, through the forest and into the sea. From white rock, boiling springs bubbled and steamed into the lake. Children were cooking megapode eggs in them. They insisted I try one and in a few minutes mine was hard-boiled. The yolk was a mottled yellow and tasted of rotten hens' eggs; perhaps this was due to the sulphurous fumes in which it was laid.

It should be a simple matter to organize the farming of these birds on a domestic basis in those areas where the pressure on their natural sites is too great. The revenue from the sale of the eggs could become a valuable asset to local villages and give them good reason to look after their birds rather than destroying them.

The distribution of the megapodes in the Pacific is very odd. They are found in northern Australia's forests and scrublands, throughout much of South-east Asia, and in New Guinea. Aeons ago it seems that they progressed down the Solomon Islands land bridge with relative ease – one species reached the New Hebrides further to the west, now known as Vanuatu – and then they apparently vanished, only to reappear on one single island in the middle of the Pacific Ocean, the Island of Niuafo'ou in the Tongan group. There *Megapodius pritchardii* lives in splendid isolation surrounded on all sides by vast expanses of ocean. To the north-east lies Samoa, to the south the remaining islands of Tonga and to the south-west the islands of the Fiji group, both large and small.

The little slate-grey bird, wings and rump brushed with rufous brown, is similar in its habits to the ones I witnessed on Simbo, spending much of its time scurrying through the under-growth on orange-yellow legs, pecking amongst the leaf litter, hoping to snatch up worms, grubs or insects in its bright yellow beak. Small crabs and snails will also fall prey to these unique creatures, of which today there are somewhere between two and four hundred. Like its relative on Simbo, this megapode lays its eggs in soil near to the warm volcanic vents which puncture the island's surface. Why should the bird be found here and not on any of the islands between Niuafo'ou and Vanuatu almost 2,000 kilometres further west? The answer lies inside caves once occupied by man.

Until very recently archaeologists who worked in the Bismarcks and the Solomons were unable to find any evidence of occupation by man older than 4,500 years. This seems odd, for man appears to have been in mainland New Guinea for at least 40,000 years; indeed some believe that agriculture originated in the highlands of New Guinea, so old are the cultures that have been discovered there. What took man so long to reach these nearest major islands? The distances were not huge by Pacific standards. Small craft must have been fashioned to reach New Guinea from Indonesia where the Papuans appear originally to have come from. What kept them on the shores of New Guinea for 35,000 years? In 1985, Jim Allen and Chris Gosden from La Trobe University in Melbourne, excavated Matenkupkum cave in New Ireland and found human artefacts 33,000 years old deep in the earth deposits. These finds are set to revolutionize theories about the movement of man into the Pacific. Who the people were and how they made their journeys is a subject for a later chapter, but one thing is clear: man was *not* held up in New Guinea, and may have moved much further east and even into the Solomon Islands far earlier than was previously thought.

Part of the reason for the inaccurate early dates is the pitiful amount of modern archaeological work that has been done in these areas. A special kind of archaeologist is required to venture into the bush and endure the harsh, sweltering conditions and the occasionally hostile interest. Many of the areas where investigation is needed are 'custom' sites, revered as ancient burial grounds, places of legend and sorcery. To tamper with these is to incur the wrath of the spirit gods, the consequences of which are unthinkable to the local people. Complex negotiations must be entered into before satisfactory permissions can be obtained from village elders, and even then rivalries between villages may result in conflicting claims of precedence. The expense and difficulty of working in these remote islands has kept their history from the world.

Tim Flannery from the National Museum in Sydney discovered charred remains and bones in the cave, indicating that the early occupants were hunters and feasted here on their prey. The beasts they captured were mainly birds: there were no bones of any mammals. This could mean that the hunters were not eating them, but if they had sufficient skill to obtain birds, they surely would have been able to hunt the slow-moving possums, tree kangaroos and rats. Perhaps these mammals did not exist in New Ireland's forests when man first arrived. If so, how did they get there? Higher up in the cave deposits, coming closer to the present day in time, possum bones begin to appear in the hunters' remains. There is another more tantalizing solution: they may have been brought in by man.

Such valuable creatures well known from the mainland would have been sorely missed in the hunters' new home. Somehow they must have been transported across the sea in the simple water craft the first colonizing peoples used and then, without competition from resident marsupials, quickly spread through the forests. That man should have played such an important role in assisting animals along the Solomon Island land bridge was a great surprise to me. Where did it stop? How many of the animals in the Pacific had in fact been moved there by man? Creatures of value for food or ceremonial occasions — brightly-coloured birds whose feathers were highly prized, like New Guinea's birds of paradise, for example — might have been candidates for a canoe trip to a new land. Was this an answer to the mystery of how some creatures reached the Pacific's distant archipelagos? It was strange to find the history of man and nature so closely intertwined.

Five months of work inside a cave on the island of Lakeba in Fiji revealed to Simon Best, a New Zealand archaeologist, that the forests there had once contained a huge pigeon which could barely fly, and megapodes. On another island, off the large island of Viti Levu, a new genus of giant megapodes, the size of small turkeys, once lived; now only their bones remain. On the island of Tikopia in the far east of the Solomon Islands archaeology has again revealed the presence of megapodes, though none are to be found on this isolated island today. This was due not to any modern influence but to the inability of the noble savage to live in harmony with his environment. The reason that there are no megapodes over such a large area in the Pacific where you might expect them is that they have been exterminated by man. The early settlers exploited their surroundings to the maximum of their potential; this fact is at odds with the notion of Paradise that generations of writers have extolled. On Simbo I pondered the thought that if traditional methods of protecting natural resources had failed, modern imposed methods would have no hope of succeeding.

On the south-western shore of Simbo, opposite the lake of Lologasa, the source of the heat for the megapodes spluttered and steamed from between huge boulders encrusted with yellow sulphur at the base of a cliff. Keti found a smouldering branch and blew smoke into the hole, which precipitated impressive amounts of steam into the air. Huge ochre-coloured boulders were scattered around where the erupted cliff had fallen, green trees clustered around the edge at the top, and at its base pandanus trees hung their rosettes of leaves over the sea. On the shore a huge tree had been washed up, its roots twisted and smooth against the sky. Had the megapodes used such logs as a raft? The monitor lizard may well have done so, if it was not brought in by man. But megapodes are strong fliers and are known to fly to offshore islands to roost at night; in fact they could have flown across smaller water gaps.

Drifting wood is a key in the colonization story. Many of the castaways that I was to encounter have ridden such rafts not just between nearby islands, but thousands of miles across the Pacific Ocean itself.

We returned around the coast to a village called Masusru and there collected a small, unshaven man called Peter Geli, with sparse greying hair and a gammy leg. He was the keeper of the *tabuna*, or skull houses. On the tip of a coral spit covered in vegetation, dominated by the high mountains behind, we entered the forest at a place known as Pa Na Ghudu. The keeper hobbled through the scrubby plants and we moved under the shade of some trees. A little further and there, clustered in piles within chests made of coral blocks, lay the darkened orbits of numerous staring skulls. There were twelve such containers, each filled with the whole and broken skulls of village ancestors carried to this sacred resting place long ago. Corrugated iron now served as a roof over most of the chambers, and as the canopy above my head rustled, the dappled sunlight danced a hideous pattern over bleached, moss-covered bone filled with decayed and splintered teeth. Amulets of shell money carved from giant clams lay strewn around. Nearby there was a small statue, a wooden torso lying on its back amongst the leaves and fallen branches, its extended earlobes projecting downwards from the sides of its head. The body was streaked with years of decay. Two large holes right through the head served as eyes. This was a *beku* idol. I felt here the mood of a darker time, when lives were cheap and skulls were for the taking as symbols of manhood. Simbo was a centre for the Cult of the Dead.

In the nineteenth century, Simbo was a major base for head-hunting raids to the islands near Gizo. Some of the earliest accounts were given by Andrew Cheyne, who visited the island in the 1840s when European traders were permanently settled here. Large numbers of heads were taken in these raids; 93 were recorded from one attack alone. What I had seen was a miniature dolmen on which the skulls were placed inside a superstructure of sticks and leaves, now replaced with the collapsed corrugated iron. The *tabuna* was an essential part of Simbo funeral rites. Of special interest were the *ungi*, widows who were compelled to commit suicide after the death of their husbands. At this time their valuables, including armrings and other shell jewellery were also destroyed. Should the skull of a particular man not be available, then a small stone monolith or *ngele* would be placed at the *tabuna* instead. The skulls were often decorated with rings of shell over the orbits, and sacrifices of food were placed in the roof over them or in a fire in front of the dolmen.

I was saddened by the atmosphere of decay and neglect here. Mike and I raised the *beku* statue upright, put some of the shell amulets in the positions they had held 56 years before, and silently left.

The return to Gizo was even worse than the journey out. A day of trade winds had left the sea confused and reluctant to let us through. It seemed much rougher in some areas than others. In particular there were two patches, each of about a mile in length, where it seemed we would be swamped. I had felt them when we crossed to Simbo that morning – it seemed an age ago. To the early navigators passing this way centuries before, such patches would have been signals in the sea. Polynesians versed in the ancient and highly skilled art of wayfinding would lie in the base of a canoe to feel the direction of swells that had bounced off islands way beyond their view. Here I could feel them too. Where the swells met, deflected from two or more islands, the sea became particularly rough. Such areas provided points of reference, made up a wave map discernible even in total darkness. I began to sense how the Polynesians and early seafaring peoples who came this way could navigate on their journeys through the Pacific. It was not difficult to imagine the tense feelings of voyagers generations before who after days, perhaps weeks, of being at sea glimpsed an unknown island ahead in the gathering dusk and knew that before them lay a new land: the fulfilment of a dream, a place to continue what the Polynesians called *vanua*, the way of the Pacific.

As darkness fell the sea took on a new menace. Gone were the brilliant sunlit crests and blue-green waters. In their place were only mountains of blackness and the roar of breaking surf. Ahead lay the lights of Gizo, the thin line of its protecting reef a streak of white in the night. There would be no protection there for us should we capsize. Our bodies would be picked up by the huge breaking seas and pounded on to the razor sharp coral heads, which would strip flesh from bone as efficiently as a whaler's flensing knife.

Chapter Three

THE FORESTS OF KING SOLOMON'S ISLANDS

Gizo

> Rainforests are one of the most complex, beautiful and important of the many eco-systems of this planet. They are also the ones that, primarily out of greed, we are destroying with the savage, unthinking ferocity of a troop of drunken apes in an art gallery.
>
> Gerald Durrell, 1986

 t is not difficult to fall in love with Gizo. The town which bears the island's name has been the administrative centre for the Western Province since 1899 and is the second largest in the Solomons, which still means it is small, numbering a few thousand inhabitants. From the wood and palm thatch terrace of the Gizo Hotel the morning after our return from Simbo I could gaze out on the gentle ebb and flow of lives here.

It is a classic tropical town. Its inhabitants drift at a languid pace, the sound of flip-flops marking their passage. Linked by long brown arms, they wander through patches of sun and shade beneath the trees which cover the main street. Their faces appear more Fijian than Papuan, most with skin so dark that a tongue is a pink flash against full black lips given to cheerful smiles. Their sparkling eyes are deep brown set in ivory. Their hair is either close-cropped and frizzy, or long and wavy, owing to a Polynesian or perhaps an early European influence. In twos and threes they stroll into town in search of vegetables from the market, beer from the store, a newspaper, tickets for boats to other islands. There is a clanging of church bells, some singing in the distance. Dogs and chickens scratch. Shrieks of laughter from behind mosquito screens mingle with the faint plucking of guitar strings and the sound of Polynesian starlings squabbling in the branches above.

A short walk past the larger stores took me to some Chinese shops, proudly decorated with green boards and yellow beading. Blue pillars supported wide tin roofs over verandas laden with smiling children who met my waves with grins and giggles. One shop stood opposite a covered warehouse in which a large wooden boat was being built. The sounds of craftsmen working with hard woods and of a Chinese overseer chattering drew me inside. The craftsmen beating at chisels with large wooden mallets were Somalis from East Africa; a legacy perhaps of the great Arab trading skill which once flourished on that coast. There seemed to be few if any power tools; one man was fitting joints with wooden pegs and glue. They told me the boat would be ready in a few months to carry trade goods between the islands for its Chinese owners.

Outside the children played in the dusty street. Through the trees along the water front, pale-blue launches clustered amongst canoes, cottons flapping in the drying sun. In the distance beyond the curve of the bay, haze shrouded an island several miles away, the great volcano of Kolombangara.

Charlie Panakera is the manager of the Gizo hotel, where Mike and I had lunch that day. He is an MP and has a Masters degree in Business Studies from New Zealand. His father was a chief from a nearby island. The Solomon Islands have a 38-member National Parliament, from which the Prime Minister is elected, and 15 cabinet members. The Queen has been represented by the Governor since full independence was granted in 1978. There are seven provinces, each with a premier and local assembly. These are largely dependent on central government for funding. In the past tourism was actively discouraged, and to date plays a relatively minor role in the economy of the Solomons, but the need for foreign exchange to develop the country will almost certainly cause it to increase in the future.

After fish, much of which goes to the USA, timber is the second largest export, principally sent to Japan and Australia as raw logs. In a few short years, this latter trade is likely to destroy many of the animals and plants which I had come to see. I was anxious to see as much of them as possible before it was too late.

In the afternoon we took a trip into the interior. The Toyota Landcruiser we drove had seen better days, but Mike's curses and demonic gear changes spurred it on to hitherto unobtainable speeds. We swept past the Gizo prison, where I was surprised to see the tall wire-mesh gates open and most of the inmates enjoying the view of the sea from the side of the road. After passing around the southern tip of the island we came to a village of pandanus-thatched huts standing a metre high on stilts. They reminded me of the Polynesian villages I had seen in the Cook Islands. Indeed, they were built by Polynesians; by Gilbertese from a nation of atolls to the north-east who had been moved here by the British Administration in the early 1950s.

Spying a small logging track, we climbed away from the coast and were soon within the forest. Much of it had been logged and was highly disturbed but there were patches in which we could glimpse its former splendour. In places I saw golden whistlers in striking yellow and black livery flitting between the branches, and small groups of olive-green birds with white-ringed eyes from which they took their name: white-eyes. Occasionally a birdwing butterfly as large as my hand would flop through the higher branches in search of flowers, and a small honey-eater probed hanging red ginger flowers for nectar. In the forest I felt at home again.

I noticed strange-looking balls looking rather like spiky grey-green hedgehogs with leaves growing out of their noses dangling from some of the trees. These represented one of the strangest of all associations between animals and plants. They were ant plants, which grow on branches high above ground, not as parasites taking sustenance from their hosts, but as epiphytes – botanical hitchhikers seeking a place in the sun. Their very lightweight seeds float and gain a foothold high in the canopy, where their leaves, once grown, can photosynthesize well. But such a lifestyle, without roots in the ground, leaves them with no food or water. They have solved this problem by getting ants to work for them.

No ant plant is more extraordinary than the football-sized *Myrmecodia*. Through small holes in its skin, tiny ants of the genus *Iridomyrmex* gain access to small purple rooms. Some of these are flat and interconnecting and here the ants rear their brood. Others are tubes, exchanging oxygen with the exterior, and some act as burial chambers for the colony's dead. These ant catacombs are covered in absorbent warts which play a vital role in providing the plant with the food it needs. During the night the tiny ants stream out of their high-rise homes down the tree trunks and on to the forest floor. There amongst the leaves, like Lilliputian undertakers, they search for the bodies of other dead insects such as beetles, aphids, or bugs that may have fallen in the constant rain of natural death from the canopy above. On finding the carcasses they begin a Herculean struggle to carry them aloft to their plant home, where they put them in the internal garbage dump to rot. Here another miracle begins. The ants seed them with fungal spores and in time these grow into minute mushrooms on which they feed. The action of the fungi also releases juices which are taken up by the wart-like rootlets and into the plant, providing it with the organic compounds which it needs. The ants get a home, the plants their daily bread: a satisfactory quid pro quo.

There are other creatures in the Solomons which have un- likely bed partners. Termites often create large colonial mounds made of particles of wood stuck together with insect saliva on the sides of trees high above the ground. The trunk and branches of the host tree are streaked with a network of tunnels leading to rotting branches and tree cores which the termites efficiently chew their way through, slowly reducing the forest to dust (wooden houses and often my notes are similarly disposed of). Termites appear like small fatty glob- ules on six legs and many animals, including humans, like to eat them, so the termites protect themselves with muscular soldiers, often armed with powerful jaws which can deliver a painful bite. This is not enough to deter other raiding insects capable of swiftly avoiding the sluggish termites, so many have additionally equipped themselves with small spray guns stick- ing out of their foreheads. From these they fire foul-smelling and extremely sticky fluids, the effect of which, if you are very small, must be rather like being sprayed on by a skunk and then covered in chewing gum.

The smallest parrots in the world are the pygmy parrots of New Guinea and the Solomons. In the Solomons *Micropsitta finschi*, the emerald pygmy parrot, and *Micropsitta bruijnii*, the mountain pygmy parrot, burrow into termite nests to rear their young. These delightful birds, barely the length of a finger and sporting green plumage, their faces adorned with pink in the mountain species and a crown of blue in the lowland species, appear to feed on lichen growing on the bark of trees, sometimes climbing head downwards to obtain it like the nuthatch of the temperate world. As many as six adults may occupy the nest while the young are being reared and none of the birds appears to suffer any ills from its creeping bedmates. The termites seal the entrances to tunnels breached when the nest holes are dug and cause the birds little offence, possibly because the parrots assume the colony odour by which termites recognize their kin. But why should they wish to nest there in the first place? It seems that the interior of a termite nest is the perfect place to incubate eggs. It maintains a high humidity and the eggs remain warm without being exposed to the harsh sun. Also, few animals wish to run the gauntlet of termite soldiers, so few predators such as rats or lizards will enter the nests.

A whoosh of wings and a loud 'quark' indicated that a hornbill was nearby, but I couldn't see it. There was a rustling of leaves. It was probably feeding on figs. A second sound like wind running through the shrouds of a sailing vessel announced the arrival of its mate, gliding in on massive black wings with a span wider than my outstretched arms. Its huge bill, topped with a casque of 'ivory', gave it a faintly ridiculous appearance. These birds are some of my favourites in the forest. They are expert at picking fruit from the branches and are also partial to nestlings, lizards and insects if they can find them. Their serrated beaks are strong enough to shear wire, but they have to toss their heads backwards to allow fruits to fall into their mouths because they have only a stub for a tongue. They are adept at catching. One I knew of, kept as a pet, enjoyed teasing the family dog by dropping food into its mouth and snatching it out again before the exasperated hound could swallow it.

There are some eighteen species of hornbill in the great forests of South-east Asia; many more are to be found in Africa. Moving east into the Pacific, this variety is filtered away to a single species in the Solomons, Blyth's hornbill, *Aceros plicatus*. It was a pair of this species that I was now watching. Its breeding habits in the Solomons are unknown, but it is likely that they are similar to those observed elsewhere in its range, which stretches as far west as Burma. Like parrots, these birds nest in tree holes high above the ground, but unlike them, the hornbill females are subjected to an extraordinary imprisonment that may last several months. Having found a suitable hole in which to make their nest the hornbills use their powerful bills to widen the entrance until it is large enough for the hen bird to squeeze in. Then the entrance is sealed up with mud, saliva and faeces until only a small slit remains, just large enough for the female's fearsome beak to protrude. Safe inside her prison, she loses most of her feathers, which serve as nesting material. Two or four eggs are laid and the family grows inside the tree, fed by the cock bird and other members of the hornbill community. Only when the young birds have fledged is the opening breached so that they can spread their wings in the humid air.

We paused on the track by a huge tree in a clearing. In the topmost branches were the most beautiful parrots I have ever seen. Female eclectus parrots are a dazzle of red and blue, while the males are emerald green, flashing red under their wings as they fly. The brilliance of their plumage is startling in the forest even at a distance. Six were engaged in a particularly raucous argument in the bare treetop. Lower down another appeared from a nest hole in the trunk and added her voice to the cacophony of sound. Soon the object of their displeasure came floating elegantly and menacingly over the crowns: a massive Sanford's eagle. As it approached on its huge brown wings, the parrots scattered in several directions, squawking through the forest, disappearing out of sight.

As we continued along the road I noticed what appeared to be a large white paper bag floating on the wind over the tree crowns. It revealed itself to be a Ducorps cockatoo, which occurs naturally only in the Solomons. There were quite a number of them in the forest. The track wound through the trees and we came suddenly on a group of hunters who had just shot one. It was still alive, blood dripping from its broken wing and splashing red on its creamy-white feathers. It cried pitifully as they proudly stretched its wings out to show off their prize.

'Meat?' I asked Mike.

'Meat,' he replied and drove on.

*　　*　　*

Since the time I first used aerial walkways suspended like flexible bridges through the rainforest canopy on the expedition Operation Drake in 1978,* this part of the forest has become the goal of an adventurous new breed of arboreal naturalist equipped with climbing ropes, micro-light aircraft and even hot-air balloons. What these explorers have been finding has astounded the scientific world and is of great relevance to the forests of the Solomons and other Pacific islands.

The sheer abundance of life in the rainforest roof came as a great shock. So many new species have been discovered that the estimate for the total number of insects in the world has been revised from one million to thirty million. Numerous creatures which conduct their whole lives without ever coming down to the ground could be studied at close quarters for the first time. A host of species formerly out of reach, a giant feast of knowledge, is waiting to be discovered, with untold benefits for modern man. In the canopy lies the answer to a crucial scientific question, the secret of the forest's natural ability to reproduce itself.

The fact that the world's tropical rainforests are being destroyed is now well known. In the time it takes you to read this page some two hundred hectares will have gone for ever. This terrifying rate applies to every minute of every day. An area the size of California disappears each year. The forests cannot be regrown in the span of a human lifetime; and they are rarely given the chance to regrow at all. In twenty years most of the world's rich lowland forests will have vanished. In their place will be infertile farmland, uneconomic cattle ranches, and a fragmented and parched scrubland populated by itinerant settlers. The long-term consequences to the planetary ecosystem are unknown; destruction on this scale has not occurred since the dinosaurs vanished from the earth. Warnings of lack of oxygen, atmospheres growing hotter, changing weather patterns and rises in sea level do little to change attitudes in the countries that own the forests, often hard-pressed for cash to fuel their ailing economies and repay interest on the international loans designed to prop them up. But the immediate

* See *The Enchanted Canopy*, Collins 1986

human and environmental effects are heart-rending to see, and they are especially acute in the islands of the South Pacific.

The mountain of Kolombangara that I had seen across the bay from Gizo is an illuminating example of the rainforest destruction business. The island is a typically cone-shaped volcano rising to 1,170 metres, and 30 kilometres across at its base. At the top mists swirl about the trees whose branches are draped in moss, sipping water from the damp atmosphere. On lower slopes there is still some forest but it has been extensively logged. The company responsible for the logging operation was Unilever, a British and Dutch multinational with huge oil-palm plantations in the Solomons. The company's timber interests were sub-contracted to the United Africa Company, which had extensive lumber operations in West Africa. Some countries there have now lost up to 90 per cent of their original forest, with dire environmental consequences. UAC created Levers Pacific Timbers to exploit the timber resources of the Solomons, and their experiences provide a fascinating case study of the difficulties of reconciling the extraction of timber with the needs of the land and local people.

In 1962, a pilot scheme had been set up on Gizo, when the Solomons were still a British colony. Operations moved in 1967 to Kolombangara, where Unilever owned a leasehold covering three-quarters of the island, most of which was only sparsely inhabited. Logging progressed anti-clockwise around the island and then along roads up the ridges, removing all valuable trees along skid tracks, leaving the rest. A fragmented forest remains. The Solomon Island Government was supposed to replant, but it has not. The business was profitable by 1970, and employed 600 people. Most of the wood was exported to Japan and Australia as raw logs for making into plywood or sawn wood; the quality was not high enough for decorative veneers, and the volume was not sufficient to be processed locally. (This could greatly have increased revenue to the Solomon Islanders, of course.)

At about this time a new religious sect was developing on the much larger nearby island of New Georgia, led by a powerful and charismatic individual known as the 'Holy Mama'. The Christian Fellowship Church, as it was known, established its headquarters at a place it called 'Paradise' in the north of the island. The Church was initially unrecognized by the Government, but successful in getting local people to pay their income into its coffers. In return for this, it built schools and redistributed the money to its vassals according to their needs. It soon became very influential. The Holy Mama invited Levers Pacific Timbers to extend its logging operation to New Georgia. Such an invitation was extremely valuable, for one of the greatest hurdles in any such operation is persuading local landowners to grant access to the land.

Despite the Holy Mama's invitation, Levers Pacific Timbers found themselves embroiled in litigation over boundary and ownership claims as they sought to define the timber rights; who should receive royalties, when and how much. The British Government failed to help, divesting itself of its responsibilities in the region as independence approached. The North New Georgia Timber Corporation (NNGTC) was created and, with the help of a Sussex University lecturer, a New Zealand forester and a Fijian lawyer, a fifteen-year deal was struck which was hailed as the 'best in the Pacific' for the local landowners. Then things began to go wrong.

The stimulus for the creation of the Christian Fellowship Church had been a vision which came to the Holy Mama under a certain coconut tree in 'Paradise'. Inadvertently this was knocked down by the loggers. The Holy Mama took it very badly, and relations soured. World opinion was also on the move. Conservationists from the Rainforest Information Centre at Lismore in New South Wales moved into the vacuum, stirring anti-multinational sentiments in the last remaining areas of forest on Kolombangara, claiming that the logged area was a desert, which it was not, though many species may have been greatly endangered, and that the landscape was scarred by landslips, which it was. The growing antagonism spilled over into New Georgia.

NNGTC continued logging at Baroa and began operations at Enoghae. Representatives of a powerful Solomon Islands family had joined the Board. They fell out with the Church, and funds no longer ended up in Church coffers but went direct to landowners. The Church lost influence and became a focus for anti-logging activists claiming local lifestyles would be destroyed and damage to the environment would result from logging, both of which were correct, and that multinationals were bad for the Solomons, which was debatable. Relations quickly worsened. Families of loggers were threatened. At daybreak on 27 March 1982 over a hundred men armed with petrol bombs attacked the camp at Enoghae. The trade store was burnt to the ground on its opening day, the employees fled into the bush, machinery was ruined. Unilever has now ceased logging the Solomon Islands.

What does all this mean to the Solomons? Their forests are not now safe. Australian, Japanese, Korean and even Chinese timber companies have been given a new lease of life through the vacuum created by Unilever's departure. Priceless trees are now being removed in a far less controlled manner. Graham Baines, Scientific Adviser to the Solomon Islands Government, told me that prior to its departure Unilever was actively doing its best to minimize damaging effects to the environment and that this was a considerable improvement on earlier practices. No such efforts were being made by other companies.

The loss of revenue to the Solomons should they stop exploiting their timber – which at present rates will be used up within fifteen years – would be considerable. Unilever's logging activities accounted for 15 per cent of the value of all the country's exports. With such influence it is not difficult for major companies to exert pressure for deals which do not always favour host countries; the latter then wake up to the fact that they have sold their best asset for a song.

Throughout the South Pacific huge areas of forest have already fallen. Many of the lowlands were cleared by the hands of the Melanesians and Polynesians themselves, making gardens to feed their burgeoning populations, but now it is commercial logging which presents the greatest threat to the wild inhabitants of the forests. Because they have evolved in such isolation and because they contain such a high proportion of unique species, the loss of these forests will be felt far more than that of an equivalent area in the great rainforests of Amazonia or South-east Asia would be. The mighty trees cladding the lower slopes of a few Pacific mountains might seem a small price to pay for the development of nations starved of funds, but the cost to their heritage and the global genetic deposit account is higher here than almost anywhere else. Once the unique natural world of one of these isolated islands with its high proportion of endemic species has been destroyed, it cannot be replaced. Something of the uniqueness of the earth will have gone for ever.

The following day I was greeted by a grinning boatman who arrived at the hotel at dawn saying that the magistrate's canoe was 'broken 'im all buggerup', which meant either that he wished to obtain yams from his garden, or that his wife was suffering from a spell placed on her by a neighbour, or that the boat engine was indeed not working. It seemed we would be unable to journey through the islands in the fearsome judicial pomp that I had anticipated. By a stroke of luck the missionary's canoe had returned and Nathan once again made himself available to us. In glorious sunshine, Mike McCoy and I set off across the ocean, leaving the island of Kolombangara to our left, and headed north-east in the direction of New Georgia to explore the myriad islets of Roviana Lagoon.

With the dark mountains of New Georgia rising before us and the turbulent waters outside Gizo left behind, we entered a scene of beautiful coral-strewn shallows interspersed with deep-water blues which led through a maze of entrancing small coral islands sprouting like tufts of green jungle from subterranean platforms. Palm trees grew thickly on most of them, and occasionally a simple thatch hut could be seen nestling amongst the trees close to the water's edge. White sand lay like a ribbon gathering up the trees. The waters here, unlike those on the way to Simbo, were as smooth as glass, protected from the Pacific swells by the outer barrier reef. Only the white streak of our canoe's wake disturbed the surface; small white breakers heaved themselves almost reluctantly on to the flat coral shelves, as if the humidity and the stillness would soon force them to give up their struggle.

There is one rough patch of water: Ferguson Passage. It was here that *PT 109*, captained by the young future President John F. Kennedy, was rammed and cut in half by the Japanese destroyer *Amagiri* on 1 August 1943. The torpedo boat sank rapidly into the shark-infested waters, leaving Kennedy and eleven other survivors to swim to a small island in Roviana Lagoon. (Today this is known as Plum Pudding Island, a name which accurately describes it and many of its neighbours as well.)

Further on we approached another small island. This too had been touched by the war, during which it was used for bombing practice. Today it is richly covered in mangrove trees and palms: a remarkable recovery. Now it is the occasional home of the sea krait, one of the most poisonous snakes in the world. It was this that I had come to visit. Ten fat fruit pigeons exploded out of the treetops with a fusillade of clapping wings as we approached; this was a safer roosting site than the main islands, where they would have to feed in the shadow of shotguns and arrows. A large osprey was quartering the lagoon. I tend to associate the osprey with the cold and Scotland, but in fact it is to be found throughout the world and is one of the largest birds of prey to visit the more distant tropical islands of the Pacific. The sight of one plunging feet first into the sea and emerging with a large, struggling, silvery fish in its talons is unforgettable.

Once we had landed we set about searching the undergrowth and stones for our deadly quarry. Weaver ants bit my toes as I probed amongst the mangrove roots, and some were carrying pieces of sloughed snakeskin, but of the possible owners we could find not a trace.

Then I lifted a large and promising stone on the beach and there underneath it was a huge sea krait, striped in black and duck-egg blue. I got such a shock that I dropped the stone back again and beckoned to Mike to come over quickly. The second time we lifted the stone the creature was not in a friendly mood. A sea krait's venom is one of the most toxic animal substances known to man. I had dropped the stone on its tail.

'Pick it up,' said Mike.

'You must be joking.'

'Quick, don't let it go.' Mike reached forward and captured the snake, grasping it with thumb and forefinger behind its head. His years of experience with reptiles showed.

'Laticauda colubrina,' he announced triumphantly, holding it aloft. 'They're very docile really. They feed mainly on eels and fish which they catch in the sea at night. Local fishermen in several parts of the Pacific swim down and pick them up from the reefs by their tails like bits of rope. Then they swim to the surface and sling them into their canoes, but nobody seems to get bitten even though the snakes slide around their feet all the way home. Most people think they don't bite because their fangs are at the back of their throats. Here, take a look at this.' He carefully allowed the snake to open its mouth and a pair of perfectly formed fangs lay exposed and well placed to bite. 'Take its head and tail and I'll get a snap of you.'

The snake hissed gently at me as I gingerly took hold of it. Its scales were unlike any I had seen on a snake before. They are smaller than usual on the belly, making the animal less mobile on land; the larger scales of its terrestrial cousins afford a better grip. Its tail is flattened like a paddle, its coloration lends it protection amongst seaweeds and corals, and it can stay submerged for up to eight hours. The sea has truly become its home.

The species we had found has spread itself all over the tropical world. The females give birth to eggs on land amongst rocks and tree roots. At this time they may cause alarm by entering villages and even attempting to eat fish that have been put out to dry. Some sea snakes have adapted even further to the marine environment and produce live young offshore. The powerful venom of sea snakes may have evolved in response to the need to immobilize fish immediately to prevent the snake from being dragged away or the prey from entering a coral crevice. The turtle-headed sea snake, *Emydocephalus annulatus*, common in Australian waters, specializes in feeding on fish eggs and has no use for fangs, teeth or venom. Other species have evolved to specialize in capturing eels, and have thin bodies from the head to the middle so that they can reach into burrows to seize their prey. In all there are about fifty species of sea snake in the world, twenty-seven of them in Indo-Australian waters. Their numbers dwindle eastwards to eight at the eastern end of Papua New Guinea, to three in Fiji and just one in the central Pacific.

In the south of the Solomon Islands, on an island called Rennell, there is a lake which is home to a very special sea snake. The lagoon, once at the atoll's centre, is now raised on limestone cliffs a hundred metres high. Twenty kilometres long, the lake contains in its brackish waters a species which is not known anywhere else in the world. Thickly growing rainforest covers virtually all the island's surface. It remains one of the least disturbed places in the Pacific and is inhabited by more endemic species than any other island in the Solomons. The Rennell shrikebill, fantail, starling and two species of white-eye are found nowhere else. At night, flying foxes emerge to forage while by day spectacular white ibis scratch like chickens around the villages. The people here are different from others in the Solomons too. They are Polynesians, and they call their snake *tugihono*.

In 1933, the yacht *Zaca* visited the island and deposited an American expedition which had Templeton Crocker as Patron. The strange snake which was subsequently collected here was named after him: *Laticauda crockeri*. Mike had been three times to collect more, on two occasions accompanied by Japanese scientists who wished to analyse the proteins in the venom of the creature, useful for pinning down its position in the taxonomic tangle which surrounds sea snakes of the world. It was important to collect the snakes because there was a threat of bauxite mining in the lake which could adversely affect the population before anything was known about it: a common story in the Pacific.

The Rennellese couldn't believe that anyone would fly hundreds of miles to reach their island, then journey for hours by tractor and trailer – the only transport – perched on sacks of taro and sweet potato, then motor fifteen kilometres up the coast by canoe, and hire carriers to transport numerous pieces of equipment for a further half-day hike across treacherous limestone *makatea* scrub, merely to collect a snake. No, there had to be another motive. The snake must be worth untold riches; the *tugihono* was being collected for sale in Japan. They must get a piece of the action. They would soon be rich. The assurance that all this effort would secure the collectors no financial reward was met with suspicion. Negotiations for access to the lake had to be handled delicately until a satisfactory payment was agreed upon.

Mike, having secured the help of children from lakeside villages, managed to collect 200 snakes and observe the habits of others beneath the lake surface using scuba gear. The *tugi-*

hono, it seems, eats a fish called *Eleotris fusca* with which it shares the lake, while a second sea snake which also lives there, the sea krait I had found on Snake Island, turns its nose up at the fish and eats eels. A mystery remains. How did the snakes get into the lake? Were they trapped there as the island rose out of the sea? Or is there a subterranean connection to the ocean? If so why should *tugihono* have evolved into a new species while the sea krait has remained so similar to those in the marine world outside Rennell? Perhaps the sea krait is a recent arrival and has not yet had time to change.

Being efficient swimmers, sea snakes had no need of land bridges to gain access to the Pacific, so the sea krait has managed to spread throughout the scattered islands, far to the west of where I now stood. Another species, *Pelamis platurus*, lives in the open ocean and does not need access to land at all. It has become the most widespread snake in the world.

The curious natural filter which has enabled some animals to spread across the Pacific while others have been confined to the homelands where they originally evolved in the west is nowhere more apparent than in the reptiles and amphibians here. In the steep mountains and forests of New Guinea there are twenty-two families of reptiles and amphibians, including long-necked tortoises and sea turtles, crocodiles, numerous lizards, brightly coloured frogs and snakes. Now that I had reached the Solomons, I would find only twelve families; the turtles and tortoises had been left behind along with limbless skinks and many of the frogs. Ahead of me lay the Fijian Islands where only eight families of reptiles are to be found, followed by the Samoan Islands with a mere four. By the time I reached Tahiti, in the Society Islands, I would encounter only two. Only the geckos, those efficient snappers of flies around tropical light bulbs, and skinks, some of the most widespread lizards of the world, have made it that far east into the Pacific.

One of the most extraordinary to succeed is the huge skink found only in the Solomons. Most skinks rarely exceed ten centimetres in length, but the prehensile-tailed skink, as it is known, is a giant averaging 35 centimetres, and portly with it. It enjoys an arboreal life, dining almost exclusively on the leaves of a vine, *Epipremnum pinnatum*, which it reaches with the aid of its muscular tail. This is curled around branches for support, a unique ability among skinks. Much of the day is spent snuggled up inside fig trees, whose numerous hollows provide perfect nest holes in which a female may give birth to her single live young. The skinks are rarely seen, tending to emerge after dark to prowl ponderously along the branches.

We continued our explorations from Roviana Lagoon, weaving through a maze of islands and channels to Boana Boana (pronounced Wana Wana) Lagoon. Beneath us mottled coral heads sped by almost close enough to touch, noddy terns picked at fragments of life at the sea surface, spinner dolphins coiled out of the sea in a cascade of spray, and a manta ray spread black wings over the deep. We arrived at the village of Munda in time for toasted sandwiches at a thatched rest-house to the accompaniment of chattering cardinal lorikeets in the palms, before pressing on to Nusambanga to obtain permission to inspect an ancient 'preserved' war canoe there. Once our negotiations with the elders were completed, and we were $15 poorer, a boy took us to see the remnants of the islanders' once great head-hunting days, rotting beneath a large thatched roof. The face of the 'Nguzu-Nguzu', inlaid with mother-of-pearl, still clung to the prow of the giant canoe at the waterline to repel 'water fiends' and protect the crew from mischievous waves and dangerous reefs where enemies could destroy them. Before raids to neighbouring islands, sacrifices were made to the 'Nguzu-Nguzu' and it would be attached to the canoe's prow, cutting the sea surface to the rhythm of a score of powerful warriors digging broad, pointed paddles into the blue swells. The canoe was twenty metres long, built of planks laced together with coconut fibre and caulked with a form of pitch. Seats of bamboo lay broken inside; decorative cowrie shells had long since fallen to the ground. This one had only been built in 1978 as part of the Independence Day Celebrations: it too was now being rapidly forgotten.

The following morning we baulked at rough weather and flew to the Seghe ferry for the journey back to Guadalcanal. From the air the islets in the lagoons below took on a special beauty, scattered across an aquamarine sea streaked with white sand and shimmering with corals just beneath the surface. The journey took us down the south-west coast of New Georgia. I could see the white hull of the *Iu-mi-nao* ferry steaming past the coast of Rendova Island to my right. The knife-edged peaks of Vangunu Island slid past, affording a glimpse of the magnificent lagoons of Marovo behind. Eastwards across a great expanse of sea, we came to the Russell Islands. From here at 7,500 feet I could see almost all the Solomons stretched out below. Neat coconut plantations, looking like collections of ordered fuzzy pinheads, covered most of the Russell Islands, a testament to the importance of Unilever to the nation's economy. A huge tanker was being loaded with coconut oil, squeezed from the copra, which would be made into margarine. Savo Island, still active and another nesting-ground for megapodes, passed below, and we flew down 'the Slot' to the Florida Islands and Tulagi Bay, scene of one of the greatest naval battles of the Second World War. Ahead lay Guadalcanal, and beyond it San Cristobal with the Santa Cruz Islands in the distance.

As far as Santa Cruz the colonization of the Pacific must have been relatively easy. The Solomons had provided an attainable bridge from New Guinea out into the ocean. For man no great navigational skills were required to reach this point. But the further east I travelled, the greater were the ocean gaps, the more treacherous the distances between island groups. What animals, I wondered, had been first to colonize new islands emerging deep in the ocean hundreds – even thousands – of miles away? I was to discover my answers in the sunken wreckages from great sea battles that had once raged here, for the first colonizers of the distant Pacific Islands travelled not across the surface of the sea, but beneath it.

In the cockpit of the plane, I noticed a small fly accompanying us on our journey towards Honiara, a colonist for a new age.

Chapter Four **COLONIZERS BENEATH THE SEA**

Guadalcanal and Honiara

> Often, standing on the shore at low tide, has
> one longed to walk on and in under the waves,
> and see it all but for a moment.
>
> Charles Kingsley, *Glaucus*, 1855

he sea, ruffled by the south-east trades, licked at the Japanese freighter *Toa-maru* as she made her way through the Gulf of Vella Lavella towards 'the Slot' under a reddening dawn sky. It was 12 June 1942. There was an air of caution aboard, but the absence of clouds promised a simple passage; a welcome dockside would follow, then the bars and the girls. But as she slipped between the steeply forested and apparently deserted islands, unfriendly eyes had been watching the *Toa-maru*.

By the time the crew saw the aircraft dropping silently out of the sky, engines screaming in a power dive, it was much too late. The first cannon shells tore fountains from the sea, then puncturing metal, a cascade of destruction through the deck. The running of feet, orders, panic, screaming, followed, then the bombs came skipping over the sea from the now low-flying fighter bombers. America had found her. Her 15,000 tons shook with explosions. Guns from her destroyer escorts hurled defiance at the sky, but they were unable to save her. The attack was over: the *Toa-maru* was doomed. Slowly, as flames curled around her hold, she began to sink deeper into the sea.

The Solomon Islands were unwittingly a turning point in the Second World War battle for supremacy in the Pacific. It was a battle by modern colonists in a world that had seen many pass that way by sea. Though their ships gave them the means to travel anywhere over the surface of the ocean, they too were forced to use the land bridges afforded them from New Guinea, the Bismarck Archipelago and the Solomons. By March 1942, the Japanese had captured Guam, the Philippines, Hong Kong, Singapore, Indonesia, Rabaul and the huge island of Bougainville to the north of New Guinea. They appeared unstoppable. Quickly they moved on to the tiny Shortland Islands off the eastern tip of Bougainville and advanced towards Tulagi, then capital of the British Protectorate of the Solomon Islands. The Resident Commissioner there, the Honourable Sir William Sydney Marchant CBE, was in a spot.

Evacuation seemed the only answer, but what effect would this have on the islanders? The idea of their 'protector' of fifty years taking flight at the 'first whiff of enemy gunpowder' was unthinkable. It would be useless to explain that France had collapsed, that the Japanese had committed treachery at Pearl Harbor, that the real defences of the Solomons lay in ruins in Singapore; the islanders were only savages after all. The enemy was apparently invincible, and could strike at any time. Was it worth defending the islands at all? A review of the defences was not encouraging. At Tulagi there was a tiny detachment of the Australian Imperial Force. Marchant was, however, of the bulldog breed. He decided to stay and fight.

The Japanese badly wanted Henderson Field on Guadalcanal so that they could leapfrog into the Western Pacific. They advanced into the Solomons and took it. The British began an extraordinary game of cat and mouse, consisting of small commando raids and waterborne attacks, often in canoes or old sailing schooners, the native crews of which would leap into the water at the first sign of an air attack and then climb aboard again to continue their work once danger had passed.

On 7 August 1942, the sky above Guadalcanal was lit up with smoke from exploding shells and the seaways were blocked with ships. The Americans landed and in a desperate battle removed the Japanese from Henderson Field. Tragically, after their cruisers and destroyers had successfully escorted the US landing, they were surprised by the might of the Japanese navy off Savo Island. America suffered one of its worst naval defeats. The wrecks of the *Canberra* and *Chicago* still lie today in Iron Bottom Sound. Numerous other ships were also destroyed. In all some 350 vessels and aircraft went down in these waters during the war. The Japanese had their comeuppance two months later when the Americans inflicted a devastating defeat on their naval forces at the Battle of Guadalcanal. They lost eleven transports and a battleship to the Americans' three destroyers. In January 1943 the Japanese withdrew their 11,000 troops by night, undetected from Guadalcanal, and so their push into the Pacific was finally halted.

The Solomon Islanders, administrative officers, planters, traders and missionaries played an important part in the war effort – not least through the remarkable system of coast watchers behind Japanese lines, which provided information on enemy positions throughout the islands. It was one of these posts which had spotted the *Toa-maru*.

Sitting with my feet propped against the sides of the dive canoe, looking across the water towards Vella Lavella, I tried to imagine the scene. The wreck of the *Toa-maru* lay some twenty metres down. Checking the air pressure and gripping my mask, I somersaulted backwards into the warm ocean. Once the silver cloud of bubbles cleared, I could just make out the green outline of the huge ship, sloping at an angle of 120° into deeper water, her bows facing the shore where her captain had tried to beach her in a desperate attempt to save cargo and lives. He had been successful. Japanese barges had removed many of the tanks, jeeps and supplies before the ship was scuttled and sank to her present position just a few miles to the north of Gizo.

Air hissed from my buoyancy vest. I allowed myself to be carried into the blue. The ocean remained bright even at depth but pressure built painfully on my eardrums; I equalized by pinching my nose. Fragments of light danced in front of my eyes: tiny phytoplankton, almost too small to see, sipping energy from the rays of bright sunlight which fractured the surface. These are converters of light, the world's greatest producers of oxygen, and their bodies are the fuel upon which the rest of the ocean depends.

Kicking with my fins I swam down through this thin soup. The huge wreck lay on her starboard side on a bed of sand, a time capsule, a modern reef. I entered through a large gash blown in her hull; not the work of American bombs but of charges laid by later bounty hunters in search of brass. I rose up through an opening in the near vertical deck, sheltered from advancing swells from the open sea to the north-east. There were few corals growing on this side of the hull. Some brightly-coloured angel fish watched curiously as my blackened fins disappeared again into the dark of the hold.

Here the gloomy interior was lit by numerous shafts of light from cracks and small openings in the hull above. I was gripped with a feeling of fear as I dropped further into the maze of twisted metalwork and unseen places where unappetizing creatures might lurk. A giant octopus? This looked like a perfect home, and what of sharks? Wouldn't these darkened caverns make an ideal lair from which to spring out on an unsuspecting diver? I battled to control my imagination. Brushing away some silt in a pool of light revealed clips of machine-gun bullets meant for American lives. A stick of four six-pound shells lay nearby and beside this a much larger object: a tank, its trackless wheels now held up to the surface, blotched with red and brown sponges which clustered like soggy fungus around a dying tree. Its gun barrel lay pointing out helplessly into the deep through the open loading bay, en-crusted with myriad life forms: a destroyer was now supporting life. A parrot fish picked morsels along the barrel: I could distinctly hear its teeth crunching against the few corals.

I dived deeper and peered into a doorway. A rusting bedstead at a crazy angle; a shower unit still in place; a cabin, perhaps once occupied by a drowning man. A crayfish peered out of the U-bend of a broken lavatory bowl. My depth gauge read 90 feet. There was no sound, only the bubbles bursting from the de-mand valve each time I exhaled. There was just enough light to see as I slowly finned my way through the long dark passage which led into the heart of the ship. The sea was much colder here, but I don't think I noticed it; I was wondering what I should do if a shark entered at the other end. A diving friend had once told me how he was trapped in a similar situation in a tunnel of coral. The passage was too narrow for the shark to turn round in. They paused and eyed each other for a few seconds, then the diver squeezed himself up against the side and the shark rushed by. I had no wish to enjoy a similar encounter.

I emerged into a greenish gloom through a doorway at the end. In the darkness I could just make out boilers and pistons, collapsed in a heap of metal at the bottom of the ship. This was the engine room. There was little light and no corals grew around the metal structures, only a green soft slime of algae. Above my head catwalks and ladders, once pathways for scurrying engineers, hung in criss-cross patterns, silhouetted against the blue-green light from openings above. Dark cryptic shadows swam across them now. Suddenly I noticed a large fin gently moving from behind a bulkhead. Its sinuous movements had me frozen. A ray? A shark's tail? Then a pair of Moorish idols emerged from the gloom, striped in black and yellow; their long white dorsal plumes had for a moment combined to appear like a menacing fin. These glorious fish have the ability to change their colours at night, dropping the yellow in favour of black. It enables them to hide better in the dark and so perhaps avoid being eaten.

Swimming upwards, following my glistening bubble train, I emerged through a hole in the deck towards the stern. A huge gantry stuck out perpendicular to the ship, and a ladder rose towards its summit. There was ample sunlight here, and a free flow of water. I felt relieved. There was much more marine life too. An immense tabletop coral and a collection of staghorns grew out from the rungs. Within their protective branches a collection of tiny fish patrolled, chasing each other from one territorial space to another in flashes of yellow, silver and iridescent blue. Small bright-blue damsel fish hovered above the surface of the table-top, ready to dive for cover the instant my alien mask approached.

These ladders and the ship's railings had been immersed just forty-five years, yet some supported coral tables over a metre across, as fine as you might see on any natural coral reef. The colonization of such artificial reefs provide some indication of the speed with which these small creatures can build a real one. I wondered if they worked equally fast on emerging islands.

I swam over the stern of the ship to the port side, the side which faced the open sea, the surface and the sun. The transformation was remarkable. Butterfly fish, bright yellow and spotted with black, and shoals of silvery surgeon fish moved across the coral landscape of lilacs, blues and yellows sculpted into impossible spikes and plates and clusters of tiny fingers. Empress angel fish, in a startling livery of brilliant orange, black and blue stripes, eyed me inquisitively. Chocolate-brown sea urchins and small bright-red shrimps picked their way beneath towering *Acropora* corals. I had to remind myself that this was not a true reef at all. So densely packed were the rounded brain corals and miniature staghorns that the original metal surface on which they lodged had vanished completely. Only the flat contours, the absence of the small cliffs and valleys that characterize a normal reef and the moderate size of the majority of the corals gave away the curious origins of this 'reef'.

Warm tropical seas provide the broth which corals need to flourish. But first they must have somewhere to place their tiny 'feet'. The surface of a coral reef is composed of millions of minute polyps, small animals resembling sea anemones. Each one is essentially a soft tube with one end fixed to the reef and the other forming a mouth surrounded by a ring of tentacles with which it gathers its food. This lowly creature is remarkable for the fact that it can secrete an external, calcareous skeleton using chemicals harvested from the sea. Into this hard shell it is able to retract and so protect itself. Its skeleton is usually connected to those of other polyps around it, so forming a colony, the architecture of which characterizes the species. The polyps stretch their tiny tentacles into the feeding waves to clutch the morsels of plankton carried in them – hence corals grow fastest on exposed coasts where the swells bring ample food and oxygen. The side of the *Toa-maru* which had faced away from the swells had little coral growth on it compared to the seaward side which received the gentle pull and mixing qualities of the ocean swells whose influence reached down from the surface above.

Polyps are fastidious in their requirements for growth. To speed the chemical processes which enable them to lay down the calcium carbonate from which their supporting skeletons are made, they must confine themselves to the warm waters of the world. In addition their tiny arms are easily clogged with too much silt, so they are not found near the mouths of rivers (fresh water would also be unsuitable for them). This is of great

importance to boatmen wishing to gain access to islands fortified by protecting reefs. Where streams enter the bay, the
run-off from rainfall carves a channel through the reef and
these breaks often provide the only access to the island through
the pounding surf.

But there is another equally demanding being with which the
tiny polyp shares its life and upon which it depends for both
food and oxygen. This is a minute alga called *Symbiodinium
microadriaticum*. It is capable of living in the open ocean as a
free-swimming organism, having two tiny tails which it uses to
swim around. The group to which it belongs is known as the
Zooxanthellae, the algae responsible for red tides which occasionally colour the oceans in explosions of growth. It is a
valuable evolutionary partnership, or symbiosis. Each cell of
the inner layer of the coral polyp possesses its own captive
alga, which uses the waste products of the polyp's metab-
olism to provide itself with the nitrogen it needs. In addition,
it obtains carbon dioxide which it uses in photosynthesis,
driven by the sun's energy, to create sugars and eventually
amino acids, the precursors of proteins. All this benefits the
alga but what of the polyp?

Inside its cells, digestive enzymes cause the alga to leak some
of the amino acids and sugars it makes, and these the polyp
seizes on to furnish its own needs. In fact 80 per cent or more of
the polyp's food requirements can be provided by its algal
lodgers. There is a cost. The polyp can only survive where
enough sunlight penetrates for its algae's chlorophyll to photosynthesize; it is therefore confined to shallow waters. This
explained why I saw no corals deep within the darkened holds
of the *Toa-maru*.

Nevertheless, this partnership has given reef-building corals
a unique advantage over other creatures as colonizers of the
tropical seas. The open waters of tropical oceans are not rich in
nutrients and so cannot support an abundance of the micro-
scopic life upon which larger creatures, higher up the food
chain, must depend. They are in effect gigantic watery deserts,
which is why they are so clear. A coral colony attempting to
establish itself on an island in such a desert would soon starve
were it not for its algal supporters which are able to generate
the energy the polyp needs from light alone. The system cannot

survive completely on its own resources, so the polyps snare 10 or 20 per cent of their needs from the water around them in the form of plankton. Because their needs are small, the meagre offerings of their surroundings are sufficient.

The parallel with tropical rainforests on land is remarkable. Here too a remarkable ecosystem designed to survive in areas of low fertility has evolved. By recycling nutrients which are often scarce in the thin soils on which the trees grow, and wasting little to the environment, these forests can survive on small inputs of 'fuel' from the sun and organic matter captured from the air. In their branches, the huge trees often support large numbers of epiphytes, which do without roots buried in the soil in order to enjoy a high-rise position in the sun. Rather like the polyps of the reef, some of these plants have been found to capture dust and soil particles blown on the wind, the plankton of the air, using fine hairs attached to their leaves. Combined with leaves dropped from trees and intercepted as they fall to the ground, these particles create an airborne compost which provides the epiphytes with their nitrogen needs. Trees too have been found to grow roots out of their branches and into their epiphyte gardens to tap the nutrient lode. The epiphytes and supporting tree branches provide an environment in which a host of creatures can live in an abundance and variety unparalleled on land, and so it is with coral reefs beneath the sea.

When coral polyps die their skeletons form the structure of the reef. Each new generation of polyps is built on the remains of previous ones. In this way these minute animals, most little bigger than a pin head, have built the largest living structures in the world such as the Great Barrier Reef of Australia as well as the numerous atolls and reefs found in the Pacific. For millions of years islands have been rising out of the Pacific, driven by volcanic eruptions, deep stirrings of the earth's molten crust. As their bubbling peaks approached the surface, warm waters bathed their underwater shores and sunlight brought colour to the rocks. Their surfaces were ready to receive the first castaways searching for a foothold in their Pacific ocean world, but if adult corals are anchored so securely to their reefs, how did they reach out across oceans to claim new islands emerging from the sea?

Had I been swimming along the coral-strewn surface of the *Toa-maru* or a true coral reef in late spring or early summer, I might have witnessed a breathtaking event. It would have been a few days after a full moon, a little after dusk. In my flashlight beam I might suddenly have seen the whole surface of the reef transformed into a gently rising snowstorm. All over its surface as far as the eye could see, minute globules coloured pink, blue or orange would be drifting from the corals. Moving close to the branch of a staghorn I would see that it was adorned with pink balls like a Christmas tree, each emerging from a polyp's mouth, each itself composed of smaller globules – in fact eggs – in a transparent envelope. Inside the capsule, a small testis is concealed which will release sperm once the capsule is released and ruptures. Examining other corals reveals capsules in all stages of release, some visible beneath the surface of their polyps, others being gently squeezed free. Regardless of species, the whole reef appears to be engaged in a massive explosion of reproduction: it is spawning. In less than a few seconds the staghorn will have released all its egg clusters; only corals release theirs in patches, but all with great speed. On rupturing, the packets of reproductive cells fill the sea with sperm. Propelled by their thrashing tails, these swim urgently in search of eggs to fertilize. Not all corals release eggs and sperm together. Some, like the mushroom corals, consist of male and female individuals, some of which will be releasing eggs, others thin spiralling clouds of sperm. To release all your eggs at one time and so swamp predators might be a valuable strategy for a coral species, but why so many different species should do it all together, creating appalling confusion for their sperm, remains a mystery. Closely related species at least do tend to spawn at different times, so reducing the danger of hybridization.

Once an egg is fertilized, a small free-swimming larva develops which joins the plankton community for a period of days or weeks before it begins to sink in the water column, hopefully to land on a suitable colonization point. While many will land on the reef where they started, others will be carried across the ocean on coastal currents and perhaps out into the Pacific. Those that survive the longest as larvae will make the finest colonists of distant islands. Again, the greater the distance to be covered, the smaller the number of species which make it, and so variety on the island reefs diminishes eastwards towards Tahiti. The cradle of much of the marine life of the South Pacific seems to have been the Indo-Australian region, for nowhere else in the world is there such an abundance of species. The reefs of Palau in the Caroline Islands and those of the Philippines offer some of the finest diving in the world.

There is one major flaw to this scheme: the currents generally flow in the opposite direction to the way the corals are supposed to have travelled into the Pacific. To the north of the Philippines lie the Mariana Islands. Here a subtropical counter-current travels from west to east across the Northern Pacific. This may have been responsible for carrying marine creatures to the Hawaiian Islands, as many of them here are closely related to those of the Marianas, and a number of the fish are also found in Japanese waters. But what of the South Sea Islands?

Millions of years ago, as Australia was separating from Antarctica and drifting northwards into warmer waters and towards its rendezvous with New Guinea, the Great Barrier Reef was forming and becoming a gigantic spawning ground for marine life. At that time the currents in the West Pacific may have been different; they appear to have spread eastwards from the Australian coast rather than northwards as they do today, and therefore could have carried corals and numerous other marine creatures towards the emerging Pacific Islands. At the equator there is also a thin strip of water known as the equatorial counter-current which flows in the opposite direction to the sea around it. This too could have carried planktonic creatures eastwards to the islands. It was not a single giant leap for the coral world; more a series of gradual footsteps, moving across the stepping stones as they arose, scattering across the western and central Pacific. These chance events alone are perhaps not enough to explain all the traffic across the ocean. Occasionally another event occurred which dramatically increased the probability of creatures from the Indo-Australian region reaching into the heart of the Pacific. It has become known as El Niño.

In the autumn of 1982 a surge of warm water raised the sea level off the coast of California by twenty-five centimetres. It then swept northwards and began to flow along the southern coast of Canada. Strange crabs began to appear on San Diego beaches, barracudas from the subtropical waters to the south prowled off ports, seabirds abandoned breeding sites, cold-water salmon vanished from traditional fishing grounds. Violent storms followed in the wake of the warm surge, destroying 10,000 homes along the coast of America. It was the beginning of an extraordinary year of catastrophic weather disruption that affected nations around the globe. The worst drought of the century destroyed stock and caused bush fires across Australia; rains failed from India to Indonesia, the Philippines and Mexico; exceptional rains sluiced through the Gulf Coast of Central America; floods and landslides took lives in Ecuador and Northern Peru. In the Pacific, Hawaiians were buffeted by a hurricane, a phenomenon seldom experienced there before, while further south the Society Islands were subjected to six hurricanes in a year – there had been none for the preceding seventy-five. On Tahiti alone 1,500 houses were destroyed. The bill for worldwide damage ran to more than $8.65 billion dollars.

The exact cause of El Niño is still something of a mystery but it is connected to changes in atmospheric pressure over the Pacific Ocean. Normally there is a high-pressure system in the eastern Pacific which, combined with the force of the earth's rotation, drives winds and therefore currents westwards towards the monsoon low-pressure system over Indonesia. At the end of 1982, for some reason which cannot be adequately explained, the low-pressure system there moved eastwards into the Central Pacific, trade winds faltered, and new westerly blowing winds replaced them. The surface current was dramatically reversed, sending a surge of warm sea 4–5 °C above normal towards the Americas. As this weather system covers a quarter of the globe, the effects were enormous.

Seventeen million seabirds which regularly nested on the giant atoll of Christmas Island in the Central Pacific abandoned their young to starvation, unable to feed them due to a lack of fish in the sea. The atoll was all but deserted by the end of 1982. On the coast of Peru the anchovy-fishing industry, formerly capable of lifting anything up to 180,000 tons of anchovies from the sea each day, collapsed. The warm water that surged eastwards capped the cool currents welling up off the Peruvian coast, and with them the vital supply of nutrients they carried to the surface from the ocean depths, on which the plankton broth, the food of the anchovies, depended.

On a geological timescale, phenomena such as El Niño, which must have occurred many times in the past, may have provided a mechanism to assist the planktonic larvae of marine colonists as well as many other larger marine creatures on their journeys to the islands. The warm surge took just sixty days to reach America, carrying with it all kinds of species which had never been found in that part of the ocean before. On this occasion, however, it was a disaster for coral reefs. Corals survive best in temperatures that do not rise above 28 °C. El Niño forced the sea to 31 °C for months. Coral reefs died off *en masse*, their algal partners unable to survive the heat. In the Galapagos Islands, 95 per cent of the corals died. Huge areas were devastated in French Polynesia and areas further west. While it may take decades for reefs to recover from these natural shocks, the space created enables other species to get a foothold on islands that would formerly have been closed to them.

Not all the larvae which make up the ocean's plankton belong to corals. There are numerous other fragments of life, the offspring of much larger and more mobile creatures – crustaceans such as crabs and shrimps, molluscs including seashells, oysters and clams, echinoderms such as starfishes and sea urchins, and of course fish. Most are invisible to the eye, yet fashioned into geometrically shaped spicules and transparent swimming appendages; they are masterpieces of marine architecture, as fragile as the finest works in glass. Many of these tiny organisms float passively in the current; some use their paddles to swim upwards at night, yet allow themselves to sink at the first sign of dawn when the intense ultraviolet light may harm them. The deep-water community includes the permanent plankton of scyphozoans such as the lion's mane jellyfish *Cyanea capillata* – whose adult tentacles have deadly stings yet provide a mobile home for certain juvenile fishes which have learned to avoid them and so gain protection – comb jellies, the deadly 'stingers' of Australian beaches, appendicularians and arrow worms, and pelagic swimming molluscs such as the violet sea snail *Janthina*, which rides on its own raft of bubbles. The plankton of reef fishes is part of the temporary plankton community, for after a period of days, weeks or even months, its members will begin to transform themselves into miniature replicas of their parents and seek out a home in which to live.

Anyone who has dived on a coral reef is at first devastated by the colours and forms of the fish they see. In fact the population of any one species may be relatively small; it is the variety which is great, a characteristic of stable ecosystems. As with the corals the greatest diversity of fish species is to be found in the western Pacific and declines to the east. On the Pacific plate, about ten per cent of the shore fish families account for almost ninety per cent of all the fishes present. It comes as a surprise that fish, which one might expect to be able to swim far and wide and so be distributed evenly around the Pacific, are often as restricted to their islands as land animals are. Some thirty per cent of Hawaii's fishes, for example, are found nowhere else. Far from being an open highway, the sea is filled with restrictive barriers to dispersal, some visible, some invisible.

A fish which has evolved to reside amongst coral crevices and graze algal growth on the outstretched arms of *Acropora* corals rarely if ever undertakes a journey any distance from its reef. The ocean depths are cold and dark, and provide little food. The huge deep-sea trenches at the margins of the Pacific plate act as barriers to the creeping dispersal of even deep-sea species. Open oceans are likewise short of supplies and may be filled with water of the wrong temperature or physiological conditions for the fish to survive. The small Pacific islands may lack the estuaries and extensive brackish mangroves some species need to act as nurseries for their young. Only pelagic species such as tuna, certain sharks, shoals of herring and Pacific mackerels are able to journey far from the islands.

To ensure the survival of their offspring, reef fish must see that their young remain close to the reef. They do this in three ways. Some attach their eggs to the bottom of the reef or hide them in nests, even dutifully guarding them against foes. Safest of all are those that brood their eggs and hatchlings inside their mouths, a habit which led many early biologists to believe that they were cannibals. Most broadcast their eggs and sperm into

The First Colonizers

Hot spots under the earth's crust have given rise to many islands in the Pacific. How animals and plants have managed to reach them, thousands of kilometres across the ocean, is one of the natural world's greatest mysteries.

1 **Coral reefs fringe an island**: Monuriki, a small island off Viti Levu, the second largest island in Fiji. Once the peaks of volcanic islands poke through the surface, young coral larvae floating in the sea descend into the shallows and begin building magnificent reefs.

2 **Coral colonies take hold**: Staghorn and fan-like agaricid corals cluster in pools on the reef flat like a collection of stone bouquets. Numerous other marine creatures, such as bright-blue damsel fish, make their homes there, forming a complex island community.

3 **Na Pali coast**: These remote cliffs on Kauai in the Hawaiian islands were first colonized by seeds borne on the waves or the wind, or even stuck to the feet of visiting birds.

4 **Insect rain**: Small insects like these crane flies, blown into the atmosphere during storms, may travel thousands of kilometres across the ocean before floating on to remote islands.

5 **Cardinal honey-eater**: A male *Myzomela cardinalis* probes a flower for nectar. Once forests were established, bird life followed, evolving into many unique forms. Ancestors of these pretty birds spread through Micronesia, and from the Solomons to Samoa. Their plumage is valued for making feather-money belts.

6 **Tongan flying foxes** (*Pteropus tonganus*): Cyclones may have blown the ancestors of these giant bats to remote Pacific islands. On wings that may span a metre they leave their daytime roosts to forage for fruit at night.

7 **Ferns take hold**: New lava flows, such as these on Savai'i in Western Samoa, trap moisture and minerals in crevices which are quickly exploited. These first plants will eventually become a forest.

8 **Crested iguana**: *Brachylophus vitiensis* is the Robinson Crusoe of the animal world. Millions of years ago, its ancestors reached the Yasawa Islands in Fiji from South America; it is not known how.

3

5

6

the sea, a dangerous exercise which may allow them to vanish on the tide. In the same way as we enjoy caviar, many fishes like to eat the eggs of their neighbours, so those species which cannot afford to guard their eggs or fill their mouths with their young face a terrible risk if their eggs remain near to the reef once they are cast into the sea to be fertilized. Some fishes have developed a remarkable instinct for the flow of currents and tides and employ a variety of carefully laid plans to avoid the legion of hungry mouths near to the reef.

Those that pair, such as parrot fish, box fish and fairy basselets, engage in careful courtship, a male signalling a female with raised fins and special colours, nuzzling her belly gently, stimulating her to spawn. The moment she is ready the female dashes towards the surface and releases her eggs in a cloud, followed closely by the male who fertilizes them with his own cloud of sperm. By doing this at dusk they minimize their exposure to daytime carnivorous fishes and crustaceans and avoid the lethal arms of coral polyps which have yet to open. Other fish, as they sense the rising tide beginning to turn, swim to breaks or passes in the reef where they know the outgoing tide flows strongly. Here fertilized eggs will be carried out to sea but return again on the flood, and over the many days that the larvae will take to mature, the ebb and flow will keep them within reach of a new home but beyond the reach of most hungry mouths. Most remarkable are those fish which use currents and eddies, the gyres, occurring in the lee of reefs, bays and islands. Fishermen know the position of these gyres and lie in wait for the fish which congregate there at spawning time. Large species such as groupers can migrate hundreds of kilometres to these spawning grounds in summertime to engage in mass-breeding congregations, producing a vast quantity of young, too numerous for predators to dent.

Many of the eggs, and the larvae into which they hatch, are carried far into the ocean and will never again see the reef where they were born. Some of these are destined to become colonizers of distant islands in the Pacific. The plankton of sweetlips and sea breams may only remain at sea for a fortnight, while that of clown fish and damsels floats on currents for three or four weeks. Groupers may drift for two months, and surgeons and trigger fish for as long as three before they metamorphose into miniature adults. Those with long plankton stages are often widespread in the Pacific.

So successful have some of the marine colonists been that they are found not only across the Pacific Ocean but all the way through the Indian Ocean as well, a fact Darwin noted as early as 1859:

> ... Many shells are as common to the eastern islands of the Pacific as to the eastern shores of Africa, in almost exactly opposite meridians.

The islands of the Pacific plate share almost three-quarters of their genera of molluscs, echinoderms, crustaceans and fishes with the Indian Ocean. But 17 per cent are restricted to the western Pacific, including Indonesia, Australia and New Guinea, and a mere 2 per cent are found just in the Pacific basin. Of the 200 species of fish to have evolved from that small percentage of genera on the plate, over half are found nowhere else in the world. It is as if, once they had reached the Pacific basin, the fishes there somehow became cut off from the rest of the world and evolved alone. Of these locally evolved fish, about half have reached islands all over the area while the remainder seem to be restricted to a curious zone running from Hawaii south-east to the Marquesas, the Society Islands, the Tuamotus and Easter Island. How could this curious pattern have developed if it was merely due to larvae floating on the ocean from west to east when most of the ocean currents do not flow that way?

A hundred million years ago, when many of the Pacific species were very different, another mechanism may have been at work. The small numbers of species which are now found only on the Pacific plate may in fact be the last remnants of creatures which were once much more widespread rather than pioneers from the Indo-Malayan region. Perhaps some form of barrier developed in the western Pacific which prevented the movement of marine creatures from the rich reefs to the islands. Some species such as the widespread cowrie *Cypraea maculifera* may in fact have spread westwards from Eastern Polynesia. Another theory is that a hundred million years ago the Caribbean plate stretched right into the Pacific before the joining of North and South America, which would explain the fact that species from this region are present in the Pacific. Additionally, islands have moved around the Pacific carrying their reef organisms with them, occasionally joining with other islands and amalgamating their fauna. Such a scenario has been proposed to explain the large number of unique fishes in Hawaii. These may have evolved after the Hawaiian Islands were invaded by fishes and other land animals from islands which began emerging some 70 million years ago to the south-east. This group, now vanished, drifted north-west and the fauna integrated with that of the Hawaiian chain, giving rise to the large number of endemic species found there today.

There are rare moments at exceptionally low tides when a coral reef lies exposed to the air. This is a perfect time to explore them, although you must take care where you place your feet. My first opportunity to see a reef in this way was in the tiny vil-lage of Baramole on the Island of Nggela in the Florida Islands to the north of Guadalcanal. Mike McCoy had suggested a visit would be worthwhile as the villagers of Baramole practise an extraordinary fishing method involving palm trees. As always in the Solomons there was the problem of transport, but Mike managed to persuade the local vet to go on a crocodile-catching expedition and to take us as passengers. Salt-water crocodiles, Mike told me, ranged here all the way from South-east Asia; some have been found as far into the Pacific as Fiji. One turned up in Pohnpei in Micronesia in 1971, 1,500 miles across the ocean from Palau where the species is normally found.

As we talked, the boring fish in Mike's tank watched us with lugubrious eyes from its position on a submerged algae-covered log. It had done this every time I had seen it, as if challenging onlookers to verify that it was in fact alive.

At half past eight, late and barely a few miles on our way to the Florida Islands on the horizon, the early-morning calm over which we had intended to travel was heaving itself into battle with the trade winds. Our driver shivered in the stern; the vet, Mike, and I were pitched in all directions. Several hours later we passed Tulagi to the north, once the capital of the Solo-mons, now a fish cannery. We eventually entered the calm Utaha Channel separating Nggela and Nggela Sule. Half-way up, a fast Japanese tuna-fish catcher came rapidly out of a creek, its high prow heading for the sea. It had no doubt rested there overnight, having deposited its catch at the cannery.

The tuna fisheries of the central and western Pacific are the island nations' most valuable resource. Skipjack, yellowfin, albacore and bigeye tuna are all fished, and blue and black marlin are becoming increasingly sought after for sport. Over 500,000 tonnes of tuna are harvested annually, of which 50 per cent is skipjack. Most of these small nations cannot afford to develop their fisheries without aid from developed countries to which they will also sell the fish. If they do not wish to be hooked on aid, they must enter into licensing agreements to foreign-owned fishing fleets at greatly reduced profit. Either way they are caught between the devil and the deep blue sea, because powerful nations such as the USA do not recognize the right of many Pacific island nations to exclude US fishing boats from the islanders' 200-mile Exclusive Economic Zone. Confiscation of an infringing vessel has led to trade embargoes by the USA. The Americans claim that the tuna is a migratory fish and cannot therefore be regarded as owned by any nation.

From the sea, Baramole village appeared to be dominated by tall palms and white sand. Sago-thatch huts lined the beach. Small pigs rooted around their stilts and the decorative shrubs with leaves mottled in red, yellow and green. Several small forested islands stood out in the bay. Cardinal lorikeets flew like miniature emperors between the palm blossoms. Children rushed to the shore to help drag the canoe over the sand. Peter, an old distinguished-looking Solomon Islander, eventually appeared and took us to his simple hut of wooden planks and pandanus thatch, raised a metre above the sand on poles. It was one of thirty or so in the village, stretched under the palms along the shore. Mike had been a friend of his for fifteen years, and we were to stay with him for a few days.

We relaxed with peanut butter on cheese biscuits and mugs of tea. The sounds of children mingled with that of aluminium pots being cleaned at a standpipe nearby. Drinking water came from a well formed from forty-gallon drums sunk on top of each other. Peter wore a tattered pair of shorts and smoked a pipe; his wife was naked to the waist and also smoked a pipe, as many older women do in the Solomons. Few people went out to fish; food on land was in ample supply, and most villagers seemed healthy. A walk in the bush revealed that much of the original forest had been cut down to grow crops, to meet the demands of an increasing population.

The following morning the first winds blowing in from the sea brought the unwelcome smell of rotting fish. The tide was low and exposed corals were dying in the sun. Stretching away from the sand, *Acropora* and staghorn corals were bared for all to see in a multitude of shapes and colours, like a collection of stone bouquets scattered across the reef. Picking my way through them, followed by curious children, I found numerous small molluscs sliding over the sand near to the shore; some grazing on the coral gardens, others extending their large siphons ahead, sniffing out their quarry – perhaps a dying fish nestling in a crevice, stranded by the receding tide. A multitude of tiny, brightly-coloured crabs scuttled before my advancing feet; miniature gobies with blue streaks and green patterns darted in the shallows. Large *Porites* corals like enormous yellow doughnuts had made a place for themselves on the reef. Their surfaces were punctured with small holes, the homes of serpulid worms which, when the tide was in, would extend their feathery gills to capture floating plankton, poking through the gnarled surface of the *Porites* like a collection of miniature Christmas trees.

Standing up and peering about, I tried to imagine what the island had been like before these creatures had arrived: a barren lump of rock recently emerged from the sea. In the floating soup that bathed its shores, algae provided a surface on which the first coral polyps must have tried to settle in a rain of small creatures carried in on tidal currents. Though countless millions would have died on the journey, enough would have arrived for colonies to begin and, once established, their spawning would have given rise to numerous others. Soon the island would have been coated with a steadily advancing reef. Once this initial habitat was in place, worms, molluscs and fishes would find food and shelter, their tiny larvae also carried in on ocean waves. Soon the host of creatures that make up the reef would be assembled, each with its own part to play in the running of this, the most complex of the world's eco-logical stages.

Continuing my exploration of the reef I noticed three species of sea cucumber resembling enormous marine slugs; one brown, another rust-coloured with white spots, and the third a dark reddish-brown like a cigar. Sea cucumbers gather minute algal detritus on their sticky feet, placing their tentacles on to the sand with great deliberation, putting them into their mouths to be sucked clean. They are related to the other echinoderms such as starfish and sea urchins. This group is remarkable in having developed an internal hydraulic system which enables the creatures to extend and retract their tube-like feet, conducting a stately progress over the ground. Many sea cucumbers also have the distinction of breathing through their bottoms, by drawing in water to bathe gills which lie concealed there. If alarmed by a passing predator or an uncaring tourist who clutches it roughly, a sea cucumber may eject from its anus a stream of sticky intestinal tubules which then rupture, spilling toxic substances which will entangle crabs and other creatures that may wish it harm. This creature has enormous powers of regeneration, as do many other echinoderms. Just as the starfish will regrow an arm if one is lost, so the sea cucumber regrows its insides after it has ejected them. Many species are highly coloured and beautiful. Some specialize in grazing over sponges, while others lead a largely sedentary life, extending finely-branched tubules into the current to trap plankton. Their simple but efficient life means that these sea cucumbers are extremely common on coral reefs, a fact which gave rise to an extensive trade in the 1800s.

The extraordinary tastes of the Chinese have led them to some curious misconceptions. They are convinced that eating the compressed hair of the rhino horn, or the reproductive organs of the crocodile, will turn them into sexual athletes. The bêche-de-mer sea cucumber, no doubt, by virtue of its shape, also endows one with irresistible charms. For whatever reason, the Chinese wish to consume it in enormous quantities. I have tried bêche-de-mer, or trepang as it is sometimes called, sliced in a soup in an obscure Chinese restaurant in Fiji, and found it perfectly disgusting, rather like lumps of glue. Nonetheless, a century and a half ago, many tonnes of the creatures were collected from the coral shores of the Pacific to be carried to the trading posts of Singapore and Hong Kong.

Bêches-de-mer were collected from the reefs at low tide,

gutted, boiled, and then smoked in huge huts made of pandanus thatch before being shipped to the Far East. The men who conducted this trade were ruthless. Native islanders were kidnapped and used as slaves. The trade continued as Fiji's main export until 1850.

Night is a time of great activity on the reef. A new shift of creatures emerges to gather food and hunt in the darkness, and those of the day seek shelter. In Honiara Mike had treated me to a night dive – my first – on the wreck of *Bonegi I*. My excitement at the thought of gliding over the reefs in darkness filled with fish I had never seen was tempered with a nagging fear of what might appear uninvited out of the darkness. Mike's preparations did little to encourage my confidence. The air-supply valve had been replaced on the tank that morning, but one of the underwater lamps would not work; nor would the spare torch. The lumochrome emergency light-stick appeared to be a dud. 'It'll be fine,' Mike said. I strode into the waves with a sinking feeling.

The wreck was one of four anonymous Japanese transport ships attacked by American aircraft and shore batteries on 15 November 1942 while offloading supplies during the Guadalcanal campaign. The four are named after a river which flows into the bay nearby; *Bonegi I* lies just off the beach. Instantly I noticed a difference in the fish patrolling its surface. Gone were the bright stripes and colours of the day; in their place sombre, large-eyed, predominantly red fish drifted by. Mike was a master of 400 dives here, and intimately knew the wreck and its inhabitants. As he led me with practised speed through the maze of twisted companionways and railings encrusted with algal reds and browns, he pointed out minute details of tiny cowries, nocturnal bivalves, spectacular nudibranches, highly coloured shell-less molluscs sporting rosettes of gently fluttering gills, banded shrimps and bright polyps extended from the coral for their nocturnal feed.

The most obvious members of the nocturnally-feeding community are the soldier fish and squirrel fish. These spiny fish emerge from the crevices and caves they occupy as schools in the daylight hours to hunt plankton over the reef surface. Their red colours are thought to make them less easy to see in the dark; most predatory fish are red-blind. In holes in the reef, a sleeping parrot fish can sometimes be seen inside a cocoon of mucous which perhaps prevents its smell escaping into the water and attracting a predator. The hiding damsels give way to silvery cardinal fish which specialize in feeding on the nocturnal surge of plankton which emerges from the crevices of the reef. The corals feed on it too, extending their tiny coloured tentacles into the sea, transforming the stony landscape. Fan corals, which may be several metres across, grow their fans at right angles to the current, holding their polyps off the reef where they must compete with others and into the tide where most food is to be found. The large amount of nocturnal plankton brings feather stars out of hiding, waving their many-branched tentacles from positions where they will be in the greatest flow of water, such as on top of a fan coral or coral outcrop. They too must filter planktonic food to survive.

As I pulled myself down a set of ladders towards the submerged deck, Mike suddenly began gesticulating that I should come towards him; he had disturbed a sleeping green turtle. It was a huge specimen, almost a metre across. Woken by our lights it began to swim in a circle, revealing two remora sucker fish like large eels attached to its belly. On some islands fishermen will capture these fish, pass a line through their tails, and throw them into the sea close to a swimming turtle. The remora swim down and attach themselves to its body so tightly that the turtle can be pulled in. I watched entranced as Mike began gently to stroke the turtle's back; it gradually became calm and appeared to fall asleep again in its original position.

There are five species of turtle that visit the Solomons; the green *Chelonia mydas* is the most common. The hawksbill *Eretmochelys imbricata* still nests on most islands, using its sharp beak to capture crabs and molluscs and to browse on certain sponges. The loggerhead *Caretta caretta* favours grapsid crabs for which it may dive to a depth of 300 metres but is a rare visitor, as is the Pacific ridley, another carnivorous turtle. The largest and most specialized species in Pacific waters is the leatherback *Dermochelys coriacea* which is rarely found near land, preferring to live in the open ocean and dine on jellyfish. Most of these species range widely throughout the world's tropical oceans and all must return to land to lay their eggs on sandy beaches. Their numbers vary greatly from region to region and most species are under threat due to exploitation by man.

Not far from where we had seen the turtle, Mike grabbed my arm. He motioned me to switch off my lamp and pointed to an area deeper down. Dimly at first, a small cloud of twinkling lights approached us out of the deep. As they came closer, the intensity of the lights increased and the cloud grew in size as they spread through the sea in front of me. It was impossible to see the source of them; they appeared to be switched on and off at will. They came still closer up the sloping beach and sur-rounded us where we crouched transfixed. I had a feeling I was being drawn into a galaxy of small stars that swirled, flowed and coiled past us before continuing up the submerged outline of the wreck. It was one of the most beautiful sights I had ever seen.

The creatures we had seen are known as lamplight fish. There are five species in the world and all are a rare sight for night divers. The genus at *Bonegi I* is known as *Anomalops*, and possesses beneath its eye an extraordinary organ which resembles a glowing elongated pearl. Within this there is a mass of small parallel tubules in which bacteria live. These produce light as a by-product of their respiration, gaining a safe nutrient supply in return from the fish. The light-producing organ can be rotated downwards and into a pocket, to switch off, or upwards to switch on, about every five seconds. So these remarkable fish can see each other in the dark, and possibly light up the plankton on which they feed.

On the beach at Baramole, moonlight was playing over the sea, the last of the day's trades were still rustling the palms, and the air still felt warm. It was time to go crocodile-catching. These reptiles are rare in mangroves and rivers of the Solomons, only occurring in any numbers in areas such as Lauri Lagoon on Guadalcanal and the islands of the Olu Malau group. They will feed on a variety of fish, frogs, crabs and birds which they search for at night on land. On the island of Aliiti Mike had seen them wait beneath trees in heavy rain to catch fruit bats sheltering from the downpour. Crocodiles have a clever technique of spinning themselves round rapidly underwater which makes escape almost impossible and occasionally twists off the limbs of prey which can then be eaten. A saltwater crocodile's jaws are particularly well designed to grab fish, having about thirty-eight long pointed teeth on the upper jaw, but cannot shear or cut very well. Hence it must store its prey beneath submerged logs or in watery larders in river banks until it rots sufficiently for pieces to be picked off at will. Having spent some time studying crocodiles on the Tana River in northern Kenya, I have a healthy respect and a lingering affection for these remarkable reptiles, which have survived as one of the few representatives of the age of the dinosaurs – despite our attempts to hunt them out of existence for handbags and shoes.

I had agreed with the others to meet them at the far end of the village; Mike would then show us the way through the mangroves to a waiting canoe which would take us to the creeks where the crocodiles were to be found. I had set off in the darkness and was now hopelessly lost. Solomon Islanders have a name for people like me who appear like bogeymen when the moon is up. They call them 'white devils'. Promises of encounters with 'white devils' after dark are used by parents to keep their children from straying at night or to confine them to the hut if some dreadful misdemeanour has been committed such as conspiring with the pigs to eat the yams. Meetings with white people are sufficiently rare on many of the islands for adults to have begun to believe in 'white devils' themselves. This would no doubt explain why any child that I cheerily approached to ask the way threw its hands up into the air and ran screaming through the village. The result was immediate. Dark faces appeared in doorways and peered out from behind palm trees. I heard mutterings of 'White devil, white devil!' I began to wonder if I was in the right village.

Eventually, I cornered a lady beside an open fire concentrating on cooking something in a pot. At my enquiry she leapt to her feet and stood terrified before me, eyes whiter than the moon. Whether or not she understood my question, she pointed vigorously at some trees to my left, and I set off towards them with relief, only to find myself, minutes later, up to my knees in a mangrove swamp. It was some time before I was finally rescued by Mike and the vet.

The best way to search for crocodiles is with head-torches: the reptiles' eyes shine out like red coals. It is even possible to estimate the size of the croc from how far apart its eyes are. On this occasion, however, we were unlucky. Mike succeeded in grounding the canoe on a large submerged log in the creek and the large specimens we had hoped to see were missing. Only their babies, less than an arm's length long, were lurking up the creeks. Mike consoled himself by attempting to catch baby archer fish, which spend their lives spitting small jets of water at insects in order to dislodge and eat them, a feat Mike thought would make them more interesting in his aquarium than the dull comatose fish which currently occupied it.

As we sat on the river bank watching the pools of light from our head-lamps reaching across the water I asked old Peter why there were so few large crocodiles in the area now. He told me that in the 1960s European hunters came to the islands and shot out all the big specimens for their skins. With a disconsolate look he muttered, 'Now big fella, 'im no more. Time befo', you cas'im one big fella too mus.'

'How you cas 'im?' I asked in my best attempt at pidgin. He grinned through his silvery whiskers.

'Hang one fella dog, on hook from tri'.' He pointed to a bough overhanging the river. 'You cas 'im no problem.' Then he pointed to a small sand spit with another tree hanging over it.

' 'Nudder time, you hang dog in tri' over some small road, with one fella barrel under 'im. When coro' cocodile jump, 'im fall in barrel.' He collapsed into chuckles, and slapped his thigh.

Early the following morning we enjoyed a breakfast of tinned mackerel, Arnots biscuits and baked yams, washed down with sweet tea. Peter inspected his old smoking frame filled with drying sea cucumbers. Despite the ideal low tide, we were told we would not be seeing the 'custom' fishing technique, which involved a specially made curtain of palm leaves 700 metres long with which to frighten fish on to the shore. Instead the whole village was going to a ceremony further up the coast to celebrate Prince Charles's birthday. My amazement was matched only by Peter's when he discovered that I had actually seen 'Haus bilong Charlie.' The festivities would take all day.

At the dancing ground, numerous islanders had gathered,

sitting cross-legged under battered umbrellas to gossip and lay out large parcels of 'pudding' they had carried in. Pigs squealed in the shadows, children chased them with sticks, dogs lay panting in the heat; for hours nothing seemd to happen at all, but the display of Solomon Island life was arresting enough. Leathery-skinned old women drew heavily on pipes and grinned toothlessly. A great deal of shouting emanated from a group of young men setting up an ancient intercom system with which to make announcements and run a raffle. Canoes arrived from the sea, bringing piles of yams which were neatly laid out for inspection. Two drunk men weaved their way aggressively through the crowd, stumbling over the neat piles of vegetables and shoving people out of the way. 'Malaitans,' muttered Peter disapprovingly.

Attempts to discover when the dancing would start were met with cries of 'soon'. Five hours later, with no ceremony in sight, we decided we had to leave to reach Honiara by nightfall. The dancing immediately began. Islanders decorated with beads, bras, feathers and grass skirts bobbed and swirled between the village huts to the beating of drums. Young men stamped and chanted under the sun.

Boarding the vet's fibre-glass canoe we said our goodbyes to a beach crowded with children, men and women. Old Peter stood waving as we rounded the edge of the bay. The channel made a deceptively smooth start to the journey before the waves stretching from across the open sea between the Florida Islands and Guadalcanal began their attack. As we soared down between the valleys of turbulent swells I wondered if cold Solomon Islanders made mistakes. Charlie the pilot sat shaking uncontrollably in the stern, spray almost obliterating him from view, expertly guiding our progress.

The sun began setting over Savo Island to our right, casting a deep red glow over Iron Bottom Sound. Another school of flying fish burst from the surf as we rode the crest of a wave, their aquamarine bodies glistening in the evening light as they flew on outstretched fins, gliding with all the grace of birds that had emerged miraculously from the sea.

By the time we approached Honiara the sun had vanished in blood into a darkening sea. Now the waves changed their tune as we surfed down them at fantastic speeds, the trick being to take them lengthways without broaching or ploughing into the wave in front. Many canoes are lost in this way. The sight of the harbour lights was never more welcome: they were pools of reassurance in the darkness.

Relaxing at last at Mike's house, I watched the baby archer fish exploring the aquarium. Its gloomy predecessor was no longer there to stare at me.

'What happened to that other fish?' I asked.

'I stuck him in with the turtle under the house.'

'Did he like it?'

'Shouldn't think so,' said Mike. 'The turtle ate him.'

Chapter Five

ISLAND CASTAWAYS

> My raft was now strong enough to bear my
> reasonable weight. My next care was what to
> load it with, and how to preserve what I laid
> upon it from the surf of the sea.
>
> Daniel Defoe, *Robinson Crusoe*, 1719

hat is it that is so attractive about being a castaway? Is it that the notion of being lost on a desert island is both romantic and dangerous? The Pacific has been home to many castaways, though most of them were not human beings and their journeys were not in boats. For all of them, to make a safe landing was merely the beginning. To be a survivor on a desert island depends on good fortune and special qualities. Far from being deserted, an island on to which a castaway is thrown may harbour hostile natives whose sole aim is a square meal. The castaway is vulnerable, having taken an unplanned journey to an unchosen destination. Landfall may bring not salvation but dangers previously unknown. The abilities to run, hide, use cunning and develop weapons all increase the castaway's chances of survival. If plentiful food and safe shelter are secured there is still one important element, over which the castaway has no control: the ability to breed.

There might be no member of the opposite sex around. The chances of another castaway of the right species, let alone the right sex, floating in on a piece of driftwood are remote. The vast majority of animal castaways reaching the shores of distant archipelagos die there without ever perpetuating their kind.

There is one other possible outcome. A female castaway may be pregnant on arrival. Should she give birth to a son, she may breed with him, but this course is also peppered with dangers.

The castaways who first reached the Pacific Islands, by virtue of their small numbers, were faced with a genetic bottleneck. Inbreeding enhances genetic defects and stifles the strength of variety. The young may share defects which become fatal in later generations. Mutation is the only hope for new lines of evolution, but by its very nature this is often unpredictable. The restrictions on the initial success of a potential colonizer are known as founder effects. They determine a plant's or animal's success and the longer-term consequences of the genetic information available to it which enables future generations to adapt and exploit their surroundings and to withstand dangers, such as strains of disease, that they may not have encountered in the land from which they came.

From the Santa Cruz Islands in the Solomons, there is no hope of moving down land bridges and drifting across short water gaps. From here the distances are very great, and natural dispersal to the Central Pacific Islands becomes increasingly difficult. It is 600 kilometres south-east to Vanuatu, 1,700 kilometres to Kiribati in the north, and more than 2,000 kilometres to Samoa due east and the high islands of Fiji to the south-east. Only those equipped with remarkable assets for survival at sea and those luck has shone upon seem to have been able to continue on the journey to the Central Pacific.

The Fijian Islands are situated in the South-west Pacific in the path of the south-east trades which, from June to October, provide them with a pleasant tropical marine climate, deluging the south-eastern coasts of the higher islands with anything up to four metres of rain a year while the opposite coasts remain parched and dry. The most remarkable animal castaways in the world live on Fiji's islands. December begins the season of hurricanes which continues through to the end of the wet season in March. There are 322 islands, of which the largest pair, Viti Levu and Vanua Levu, make up 87 per cent of the total land area of 18,330 square kilometres. The group is further divided into 'high' and 'low' islands, the latter being principally atolls or raised coral platforms. The 'high' islands are of volcanic origin and have been in existence for about 50 million years. This is a relatively short time in geological terms, but nevertheless makes them much older than the majority of islands in the Pacific, perhaps remnants of a sunken continent.

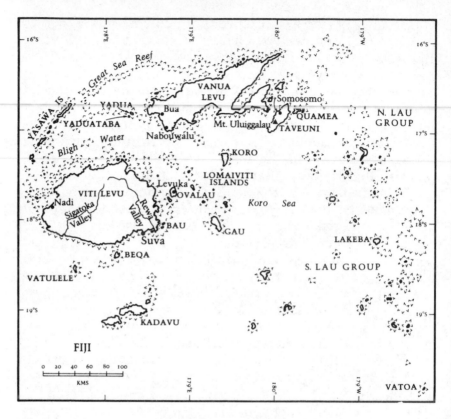

Their interiors are extremely rugged with majestic grey peaks rising to 1,200 metres on Viti Levu, often covered in a scatter of scrubby green vegetation. The high rainfall supports thick tropical forest on the eastern slopes of the high islands while huge areas of yellow-green 'talasiga' grassland dominate the drier parts, where the original scrubland was cleared for crops. Huge barrier reefs fringe these magnificent islands; the Great Sea Reef off the northern coast of Vanua Levu is one of the largest in the world. From May to August humpback whales pass through the islands with their newly born calves.

Almost half of Fiji is still forested. The variety of plants is far richer than that to be found in the islands of Polynesia. The forests are the key to the diversity of island life here. Almost half the native species are endemic to Fiji; most of these are to be found in the rich forested areas which remain, such as the Sigatoka valley on Viti Levu. Orchids and ferns adorn the branches of trees in the moist uplands. Ten unique palm species are confined to south-east Viti Levu.

There are no large native mammals in Fiji. The possums and giant rats of the Solomons have been left behind. Only bats have been able to cross the ocean to the islands. Two of them are unique to Fiji, including the extraordinary monkey-faced fruit bat *Pterolopex acrodonta*. Nothing is yet known about its ecology; this is true of many of the animals in Fiji. In addition there are beautiful parrots and lorikeets, the most highly coloured pigeons in the world, fantails, kingfishers and such delights as the pink-billed parrot finch and the long-legged warbler. Forty per cent of the sixty land bird species are endemic to these islands, as are eight reptiles. Two frogs managed to colonize the islands and occur nowhere else in the world. Some 3,500 species of insect also share the islands. Half the larger moths and butterflies – inexplicably – cannot be found on islands to the east or west. The largest insect in the world, a huge longhorn beetle whose legs would overlap a dinner plate, also lives here on a single island and nowhere else.

What accident of nature was it that enabled these creatures to reach Fiji across the huge water gap that separates it from the Solomons? What special characteristics do they have which enabled them to survive here and why are so many of them not found anywhere else? A walk along the seashore reveals some interesting clues. The most obvious is perhaps the coconut. This tree produces a remarkable seed: it floats, and contains plenty of fluid with which to nurture its offspring together with a large supply of food. It would seem ideally suited to drift across the surface of the sea, but can only survive for a period of weeks. It has managed to spread itself around the coasts of many tropical regions as well as the islands of the West Pacific Polynesians may have introduced the coconut to islands further east and to the Hawaiian Islands. The coco de mer of the Seychelles was originally known only from the twin coconut-shaped seeds that were washed up upon the shores of the Indian Ocean. As no trees could be found, these were believed to belong to a tree that grew in the ocean depths. Not until the islands were discovered and tall palms bearing the seeds were seen was the mystery of their origin solved.

Wandering along an undisturbed beach front I would often

find pandanus trees with their scratchy ringed stems and palm-like leaves, *Casuarina* trees standing on dark, rock-hard trunks, their wispy leaves waving like hair on the wind. Morning glory creepers clutched at the sand, opening pale-pink blossoms to the sun. All of these plants have seeds which float, though those of the pandanus do not float very well. These and the seeds of many other plants lie cast along the strand line, just as they might have been millions of years before when they first arrived on the deserted shore of a newly emerged island. In some cases it may not have been a seed but a whole tree or branch which made the journey. To colonize, seeds or perhaps rosettes of leaves which can sprout roots must also resist the drying effects of salt water for weeks; they must land alive. After storms, trees and branches wrenched from banks by flood waters occasionally form into rafts of vegetation which may remain entwined for many miles out into the oceans. These too provided a means of sea transport for the first colonizing plants.

Though the prevailing currents flow in the opposite direction, most plants in the South Pacific, as in the case of the marine organisms, have their closest relatives in the Indo-Malay region and Australia. Perhaps the equatorial countercurrent was sufficient to carry them east, though it flows many miles to the north of Fiji. Storms and random events such as El Niño may also have helped. Once on the shore a plant castaway must establish itself. A large wave may throw a *Casuarina* seed high on to the rocks where its roots can gain sustenance from the rich volcanic soils beneath. Resistance to salt air will work in the seedling's favour, but the climate may eventually be too dry for a tree. Equally a special insect may not yet have arrived to pollinate its flowers, or there may be no other plant of the same species from which to receive pollen. Then the ability to self-pollinate would be an advantage; such plants are often the first successful colonizers.

The most successful plant castaways tend to be weedy. They grow vigorously in marginal habitats such as rocky scree slopes, new lava flows, or beaches. The dark rocks of a newly formed island offer numerous chances to the pioneering plant. Crevices provide shade and moisture; volcanic soil, though sparse at first, is rich in untapped minerals for growth. Small ferns can be seen sprouting in the shade of boulders on Hawaii's lava fields a matter of weeks after the rocks have been spewed molten from the earth. The success of one plant favours others whose seeds may grow in its shade, and quench their thirst from rain which trickles down their benefactor's stem or trunk. The process of colonization speeds up once a vegetation cover grows, providing further shade from the harsh sun and retaining precious moisture. A new eruption may bring catastrophe, burying small plants and trees alike in layers of ash or pumice. The *Metrosideros* tree, a species of the myrtle family common to many Pacific islands, has special aerial roots which grow upwards in such circumstances and prevent suffocation, helping it to survive where others have died.

Not all plants arrive by sea. A mere 14 per cent of Hawaii's native flowering plants are adapted to ocean dispersal. Darwin showed how soil scraped from the legs of birds could sprout plants from concealed seeds. Many seeds no doubt arrived in this way, through birds blown off course, perhaps arriving on an initially barren and lifeless island. Rain would soon wash any seeds from a dead bird's feet to sprout and perhaps provide food for the next bird to pass that way. Wind too played its part, carrying lightweight seeds across great distances in storms to islands far out in the ocean.

That this process works is evident from the rapid colonization of the new islands which emerged in human times. Greatest of these was Krakatoa, which destroyed itself in a titanic explosion in 1883. All the vegetation and animals on the central cone which remained were roasted, yet within weeks numerous plants had arrived, although the nearest land, the northern coast of Java, is thirty kilometres away. A few years after the eruption a python was living on the island. The sequence is remarkably consistent. A researcher observing the colonization of the central cone of a volcano off the coast of New Guinea made a careful record of every plant that colonized the barren island, only to have all his plots destroyed in a second eruption. Diligently, he recorded them again and to his amazement found the plants had recolonized themselves in exactly the same order.

The University of the South Pacific is based in Suva. USP, as it is known, is one of the few places in the Pacific to which Melanesians and Polynesians can go to take degrees and where serious study of the Fijian Islands and their wildlife is carried out. Dr John Gibbons was to some extent an academic castaway who had found refuge in Fiji. He had three great loves: a good argument; his wife and family; amphibians and reptiles. The first had almost cost him his job at the University, where he was noted as a somewhat eccentric and troublesome individual. The last, and his views on how amphibians and reptiles had reached the Fijian Islands, had brought him notoriety in the world of science. I had had the good fortune to meet him on my first visit to Fiji at the end of my voyage aboard the Scientific Exploration Society's research vessel, the *Sir Walter Raleigh*, the previous year.

'Now this one is really fascinating,' he said, peering into a large glass aquarium in the porch of his home in Suva, the capital of Fiji. It was apparently filled with soil and a few stones. 'It's our only poisonous snake. Most people don't believe there's a poisonous snake on Fiji, not even the locals. I know it's in here somewhere.' He began to scrape away at the soil with his finger and eventually dug out a small glossy brown snake about the size of a thin pencil.

'*Ogmodon!*' he announced triumphantly, passing it to me.

'Doesn't look very fearsome to me,' I said, examining its small head which blended smoothly with its body, no doubt aiding its passage through the soil in which it lived. 'Actually it's a cobra, a member of the Elapidae. It's got tiny hollow fangs with which it injects a nerve poison, same as all the elapids in Australia. Most of the group seems to have evolved there. It doesn't bite humans though – not yet, anyway,' he chuckled.

The *bolo*, as it is known to Fijians, is extremely rare. It is only found on the island of Viti Levu, although there are some larger soil-burrowing snakes in the Solomons. It appears to be nocturnal and to feed on soil insects and worms. More than that is speculation. John Gibbons was trying to breed it. There is one other much larger snake in the Fijian Islands, the Pacific boa *Candoia bibroni*. This snake may grow to a length of three metres and as thick as a woman's arm. It varies greatly in colour, a fact noted by Fijians, who regard the redder forms as being more aggressive than those of an olive or yellow-brown colour. It is as much at home in the trees as on the ground and preys on rats, mice, lizards, birds and even bats. *Candoia* injects no poison but, like the pythons to which it is related, crushes its victims to death.

'Do you fancy going on a frog hunt?' John said as we relaxed over a beer. 'There are just two here, nothing like the giant ones in the Solomons but nonetheless very interesting.' With Mike McCoy I had found giant frogs the size of footballs inside caves in Guadalcanal. In the darkness they made prodigious leaps through our torch beams, landing with dreadful plops like large water-filled plastic bags. But these monsters are unrelated to Fiji's frogs and have not managed to cross the ocean to join them from the Solomons. Through John Gibbons I was able to discover the reason why.

'Frogs die quickly in sea water, so how Fiji's frogs got here is a bit of a mystery. The largest is the Fiji ground frog, which eats everything including its own young. It leaps on them and then stuffs them into its huge mouth with its feet, a true Fijian cannibal. The other makes nests up in the trees. We might find some at Tholo-i-Suva forest park.'

The following day, after a drive of about ten kilometres through the suburbs of Suva, we left the car at the Forestry Station and wound down a series of steep slippery paths through thick forest, the remains of an area that had been selectively logged many years before. It began to rain heavily, and John's sense of direction was not one of his greatest attributes. After numerous changes of route we finally found ourselves scrabbling along the boulders of a stream bed and peering into the bases of large fan-like pandanus trees. With boyish enthusiasm, John prised the huge leaves apart. We were completely soaked before a small brown frog, about half the length of my finger and spotted in white, hopped out of a mulch of decaying leaves and sat eyeing us from a leaf.

'*Platymantis vitiensis*,' intoned John with an air of reverence. 'This one's a bit dull. Some are a lovely yellowish orange, with a pale line down the spine and thigh; others have white or yellow blotches on them. In fact their colour is very variable, though I don't know why. They don't have too many predators here so perhaps there's no reason to be camouflaged. You can find them on the two big islands and a number of the others which have good forest left.'

I could see that it was a tree frog by the adhesive pads on each of its toes which enabled it to cling to leaves and slippery surfaces in its arboreal world. John held it out towards me.

'The ground frog is about twice this size, and a better swimmer, though this one is the better jumper.' With that it leapt out of his hand on to my arm and then into outer space. A frantic pursuit followed during which it was finally cornered in the process of trying to make its escape between two large boulders.

'This is a marvellous frog because of the way it breeds. As you know, most frogs require water in which to lay their eggs, which then hatch into free-swimming tadpoles which metamorphose into froglets. This fellow does not. It lays its eggs up there in the axils of the palm leaves where it's nice and moist. The eggs are enormous but there's no need for water. The froglets hatch direct from the egg without a tadpole stage. The ground frogs do the same.'

This form of direct development is common in tropical frogs. It may explain why the two Fijian species are the only amphibian castaways to have reached this far into the Pacific. An adult frog in the sea will die quickly but its eggs are more resistant and could have survived in the leaf bases of floating pandanus trees washed out to sea in storms. Some may have landed on the shores of Fiji and young froglets would have emerged. For a frog requiring water for its tadpoles, the problems would just be beginning, because oceanic islands are often extremely dry. The Fijian tree frog may have succeeded where others failed because it had only to keep its eggs moist under leaves to spawn a new generation. Even so the possibility of these frogs or their eggs surviving a long sea journey is a remote one. It may be that the larger ground frog was brought to the island by Polynesians for food; frogs' legs are considered a delicacy by Fijians as well as by the French. However, the fact that the Fijian ground frog is different from all other frogs suggests that it has been on the island long enough to evolve into a new species. Evolution does not move within the time span of human lives. It takes much longer, and longer than the 3,500 years that humans are supposed to have been on Fiji; unless of course they in fact arrived much earlier.

The distribution of frogs in the Pacific Islands has long been argued over. There seems to be one simple rule. The further away one gets from a continental landmass, the smaller will be the proportion of frogs with 'aquatic' eggs and the larger the proportion with 'terrestrial' eggs. This is also true in the Caribbean, where the *Eleutherodactylus* frogs, which have terrestrial eggs, have managed to colonize the islands from Central America. One of the few species to have made it that does require water for its tadpoles lays its eggs in arboreal water pools, inside air plants known as bromeliads which grow in the branches of trees on the islands there.

The *Platymantis* frogs originated in the oriental region and can be found all the way from the Philippines to Fiji. It seems that there was once an island chain, now largely absorbed into the northern coast of Papua New Guinea, which may have enabled these frogs and many species of lizard to island-hop through New Britain and the Solomon Islands with just a few making it to Fiji. At this time Australia had not yet moved northwards from the ancient supercontinent of Gondwanaland to collide with New Guinea, pushing it into this island chain. That may explain why there are nine species of *Platymantis* in New Britain and the Solomons but only two in New Guinea. When New Guinea was pushed northwards, New Britain and the Solomons were invaded for a second time by reptiles and amphibians with origins on Gondwanaland. There are some 200 species of frog in New Guinea, but few ever colonized the Solomons. The reason is that most of them require water for their tadpoles and, though many may have floated across the sea from New Guinea to islands further east, only a few were able to find fresh water in which to breed before they collapsed and died. Only *Rana* and *Litoria* frogs, including the giant I had found with Mike McCoy in a Solomon cave, are Papuan in origin. Their need for fresh streams and pools in which to raise their young prevented them from gaining a foothold in Fiji.

In the valleys of Tholo-i-Suva, grey mists were slowly lifting from the undulating canopy of trees. The rain had eased and, as if to welcome the respite, large butterflies began to float between the flowers. A flock of beautiful collared lorikeets screeched their way between the crowns on emerald-green wings, flashing their bright-red breasts. They wheeled and turned in the gathering sun and finally settled in a large tree in blossom not too far away. Like a collection of marionettes they swung and scrambled among the flimsiest of branches, reaching for the nectar and pollen inside the blooms with their long fur-tipped tongues. One pair hung upside-down; each on one leg grasped its opposite's free claw as if to shake hands, one nuzzling the purple-blue neck feathers of the other with a bright orange-tipped bill. Suddenly they fell away squawking and chattering, tumbling all the way to the ground in a flurry of wings. At the last minute they broke free and soared into the air, followed by the rest of the flock squeaking and squealing their way through the valleys in search of a new crown of flowers.

A distant 'woof', like that of a hound, indicated that a barking pigeon was nearby. I picked it out with binoculars: it was feeding on the large fruits of a tree, plucking them from the branches and swallowing them with apparent ease. It was a large smoke-grey pigeon with an olive-coloured back, its plump breast almost a salmon pink. The seeds it had eaten would later be scattered through the forest, perpetuating the tree species upon which it depended for food. It flew to another branch, allowing us a glimpse of the rust colour under its wings which distinguishes it from the equally large and similar-looking Pacific pigeon, whose underwing is dark. As the barking pigeon landed, it bowed and flicked its tail. It had a majestic air; aptly so, as it and the Pacific pigeon belong to a group of birds known as imperial pigeons. The Pacific pigeon is found from New Guinea to western Polynesia, but only in Fiji will you hear the pigeon that barks like a hound.

In walks through the forest there were always several kinds of lizard to be seen, scurrying over the leaves and branches. They are inquisitive, and sitting in a patch of shade I needed only to wait a few minutes before a number of them would appear to sunbathe or watch my every move from behind a large dried leaf. Some were jet black and as long as three of my fingers; others were smaller and sported iridescent blue tails which they would thrash from side to side like coils of electricity should another of their kind approach. All these lizards are known as skinks, and the *Emoia* skinks have been extremely successful in colonizing the Pacific. Some such as *Emoia cambellii* live amongst the treetops and breed inside ant plants, where the moist environment helps to keep their eggs at just the right humidity for incubation. John Gibbons discovered them while he was rowing around the forest canopy in his boat, when the closing of the Monsavu dam in the centre of Viti Levu had flooded the trees.

Reptiles are much better castaways than frogs. They have tough skins which can better resist sea water, and they can do without food for weeks – or even months if they have a special store of fat, as do some geckos. Their eggs are also hard-shelled and better proof against the sea. Despite having no wings to fly, lizards are supreme colonizers and have spread far more widely in the Pacific than frogs, freshwater fishes or mammals.

Geckos have colonized an extraordinary number of tropical Pacific islands from Micronesia throughout Polynesia. There is hardly a single place where the familiar 'tchuck-tchuck-tchuck' sound will not be heard from behind a picture or from the rafters of a hut. For some inexplicable reason geckos engender a reaction in Pacific Islanders similar to that of a terrified housewife who leaps on to a stool when she sees a mouse. Fijians are convinced that the gecko is highly poisonous, that its bite will result in almost certain death, that it will cause leprosy should it shuttle across your face, that it can poison a man's food or drink. Normally calm Fijian *maramas* have been known to take leave of their senses at the mere thought of a gecko tangled in their hair. None of the several hundred species in the islands in fact do this, though the appearance of the mourning gecko creeping about the walls on its padded toes, peering about itself with huge lugubrious eyes (its scientific name is in fact *Lepidodactylus lugubris*), is enough to make anyone believe that it is up to something sinister.

There is one species which has a deserved reputation: the voracious gecko, *Gehyra vorax*. This species grows to the size of a rat, and if disturbed by a probing hand in its forest home clamps its jaws bloodily on to exposed flesh and will refuse to let go – until death, so Fijians believe. They are therefore quite likely to shout and leap about the forest with the poor gecko hanging on for all it is worth and eventually resort to beating it with sticks, which only forces it to bite harder. It will refuse to let go even after it is dead, though alive it will do so quite naturally after about ten minutes if left in peace.

Those geckos that live closely with man, often seen snapping at flies around light bulbs, may well have hitched a ride with early Polynesians in their canoes, concealed amongst the pandanus mats or crop plants stored for the journey. Many of the most widespread geckos in the Pacific are capable of a minor miracle of nature: they are parthenogenetic, producing young without the need of a male to fertilize them. A single lizard or even a single egg thrown on to an island could there-fore start a whole population. A species requiring a male and a female to arrive simultaneously would have far less chance of success.

Skinks often creep under the bark of trees and lay their eggs there, a habit which would have assisted their passage across the oceans to many islands all over the Pacific, from forested volcanoes to the smallest coral islets. The emoids seem to be recent arrivals in the Pacific, following the collision between Australia and New Guinea. In contrast *Leiolopisma* skinks arrived much earlier. They are widespread, but only in scat-tered pockets; on just one island in the Lau group in Fiji, for instance. This species has relations in Australia, New Zealand and New Caledonia and as far away as Mauritius in the Indian Ocean, so may be a relict from an earlier Gondwanaland age.

As we trudged up the track and out of the valley I could not help wondering at the odds against island castaways ever making a successful voyage. Somehow it all seemed incredible – a tiny soft-skinned frog bravely clinging to a pandanus tree, guarding its little clutch of eggs. My canoe journeys through the Solomon seas had shown me what little chance there would be for the survival of even quite sturdy animals clinging to rafts of driftwood. The distance from the Solomons to Fiji is 2,000 kilometres, and the winds and currents are generally in the wrong direction. The frog would keep ending up back where it started. Yet I had seen one whose ancestors had made just such a journey, many hundreds of thousands of years before. How could they have done it?

There is one factor in the colonization of the Western Pacific islands which may be of supreme importance but until recently was almost completely ignored – the Ice Age. 18,000 years ago the world was in deep freeze. The polar caps extended towards the equator in vast sheets of ice many miles thick, grinding all that lay before them. Though New Guinea at one time had its own glacial ice cap, little of it remains today. Only at 5,000 metres on Mount Jaya is snow still to be found. The reserve which includes the mountain and reaches to the coast must be the only place in the world where it is possible to visit a glacier and a coral reef in the same National Park.

So much water was frozen into the ice caps that the sea level was dramatically lowered, at its maximum by as much as 100 metres or more. Consider the effects of even a ten-metre drop in areas where there is little tide. The sea was drained to such an extent that the coastline of the Indonesian, Australian and Papuan land masses was dramatically changed. It was possible to walk from Singapore to Sumatra, from Java to Borneo, from Cape York to Port Moresby. As we have seen this proved of great assistance to many creatures migrating eastwards to New Guinea from the Oriental region, but the effects went much further, out into the Pacific itself. A maritime map of the Western Pacific islands reveals that if the surface of the ocean dropped by 100 metres, vast numbers of islands would appear out of the sea, for in between the isolated groups that are visible today are numerous others hidden just beneath the surface, remnants of former lands now sunk or eroded away. The Solomons, a collection of seven main islands today, would gradually emerge as a huge single mountain range over 600 kilometres long, linking Bougainville, Choiseul and Isabel. Further west, as the water drained from the coral reefs fringing the island nations we know at present, a great chain of new islands would emerge as stepping stones across these from Santa Cruz towards Tuvalu, on to the Wallis and Futuna Islands, eventually almost reaching Samoa. In the ocean between Australia and the ancient island of New Caledonia, a massive new land would begin to rise from the waves, almost as large as New Caledonia itself. On either side many smaller islands would pepper the sea, linking what is now Queensland to New Caledonia and the islands of Vanuatu further northwest. Fiji would still be isolated by great expanses of ocean to the west, but to the north there would be land much closer. To the east, Tonga, today a kingdom of isolated atolls and volcanic islets, would present a dense arc of islands which could capture traffic moving east.

The emergence of these mysterious islands was not a one-off

event. The surface of the Pacific, and indeed of oceans all over the world, has been moving up and down in this way for perhaps two million years, in concert with the most recent cycle of ice ages. Further back in time it may be that other ice ages had a similar effect. The Fijian frogs, unwillingly crossing the waters of the Pacific, would have found their journeys much shorter then. In many cases the original routes and islands over which frogs could have hopped or reptiles could have crawled have long since been submerged, their collections of creatures with them.

The exception always proves fascinating, and such a one is to be found in Fiji. This beautiful creature did not island-hop or raft across the sea gaps from South-east Asia but appears to have crossed an impossibly long stretch of ocean from America. It is an iguana unlike any other in the world and has confounded all attempts by biologists to explain its existence. I first encountered one in John Gibbons' living room.

'This is William.' John prised the large iguana from the curtains where he was making a ponderous bid for the safety of the runners at the top. His enormous claws enabled him to cling determinedly, but John deftly unhooked them and brought him over.

He was beautiful. His body was a deep emerald green with three narrow white bands bordered in black down each side. On his shoulders there was a prominent crest of black spines which diminished towards his long tail. His nostrils were picked out in bright yellow and his eyes were the colour of gold. Arthritis had left his head bent at a most peculiar angle, making him look rather like an aged colonel greeting someone he couldn't quite place. William was a crested iguana, *Brachylophus vitiensis*, a species unique to the Yasawa Islands to the north-west of Fiji and only discovered to be living there as recently as 1978, by John.

'I only found them by a complete accident. I was visiting an island called Yadua off the coast of Vanua Levu. The locals told me of another island nearby with some iguanas on it. There's a similar iguana on a number of islands in Fiji and I expected these to be the same. So I went there to take a look and found a totally new species. You don't usually find new vertebrates as big as this these days.'

John's discovery proved that there were two *Brachylophus* iguanas in Fiji. The second species, which has pale blueish bands, had been discovered earlier, is much more widespread in the islands, and has also been found in Tonga. The island of Yaduataba, which contains the last sizeable population of the much rarer and more spectacular crested iguana, became Fiji's first wildlife reserve in 1980.

'The big question is' – John leaned forward as though about to confide some great secret – 'what on earth are they doing in Fiji at all?'

These iguanas are one of the great enigmas of the biological world. Unlike the majority of animals and plants in the Pacific they have no close relatives in the Oriental region, but appear to have evolved in the Americas. Their closest relatives are not the remarkable marine iguanas of the Galapagos Islands, which spend their lives scrambling over the harsh volcanic rocks there, occasionally plunging beneath the sea to crop seaweed growing under the surface, nor the land-living species which also lives there. The crested iguana's origins appear to be even further east, in the West Indies. How on earth did the Fijian iguana make the journey over 12,000 kilometres from the Caribbean to the Yasawas?

Iguanas turn up in other strange parts of the world. There are some in Madagascar, but none in Africa, though fossils show that they used to be there. Millions of years ago, when Africa was still joined to South America, iguanas roamed the super-continent. Madagascar was then still attached to the east coast of Africa. As Africa and America split, and the Atlantic grew, Madagascar split off into the Indian Ocean, carrying with it the very special collection of creatures found there today, including the extraordinary lemurs and some iguanas. On the mainland of Africa, the iguanas were extinguished by a new race of more successful lizards, the agamids, but those on Madagascar were protected by the sea and so have survived. 'But that still doesn't explain their presence in Fiji,' I exclaimed.

'Exactly,' said John, peering at William, who appeared to have nodded off, 'I'm not sure he's awfully well.' With that he scooped him up and galvanized him with a piece of paw-paw before putting him on the curtains. William climbed eagerly up and sat eyeing us from behind the pelmet.

'I think they rafted here all the way here from the Caribbean,' John announced, swinging round.

'But Central America's in the way; there's no way through.'

'Ah, yes, but five to seven million years ago, there was a huge sea gap. Central America was underwater. They could have floated across.'

'Why aren't they found on any of the islands between here and South America then?' I countered.

'Well, there aren't that many, and they may not have been there when the iguanas made the crossing. The Marquesas are young; so are Tahiti and its islands. The currents tend to go north of there anyway. Most of the Cooks have been around for less than ten million years, and Rarotonga is only two million years old. The next stop is Tonga, where the iguanas first occur, but some of Tonga's islands are as much as forty million years old, almost as old as Fiji. It may be that the Tongans took them there. They were always moving things round the place – great seafarers.'

'What about Samoa? It's only a little way north-east.'

'Less than two and a half million years old.' John looked triumphant. 'A chum of mine dropped a bottle in the sea at Galapagos and it fetched up here a year later. The journey could take as little as five months.'

'That's one hell of a long time for an iguana to be sitting about on the ocean,' I protested.

'Yes, but these fellows have got some very special tricks up their sleeves. They survive almost entirely on a diet of leaves and could eat the foliage of a tree they had been washed out to sea on. They sneeze salt through their noses, and their eggs have the longest incubation period of any iguana in the world. You should see for yourself. Take a trip to the islands. Biren Singh at the National Trust will help.'

It was not until the following year that I was able to take up this suggestion and by that time tragedy had struck. Returning from an island in a small dinghy to a ship anchored outside the reef, John's craft was overturned in huge waves. He and all his family were drowned. For expatriates, grown complacent about the challenges of the Fijian coasts, it was a salutary warning. The dangers of the tropical sea should never be taken lightly. It was with more than a moment's sadness that I returned to Fiji, knowing that John's career had been cut short at a time when his work on the colonization of the Pacific, though often controversial, was set to break exciting new ground. I decided to take up his suggestion and try to see the extraordinary iguanas.

Visiting the world's greatest castaway was a complicated process. First permission had to be obtained from Biren Singh at the National Trust. Beleaguered by mountains of paper, starved of funds, governed by a somewhat spineless council which gave him no authority to do anything, Biren Singh was charged with protecting everything from Fiji's fine buildings to its natural landscape and the animals in it. It was a hopeless task and he was glad of my offer to transport huge signboards to the island declaring it a nature reserve – seven years after it had been formed. There were copious letters of introduction to the various chiefs and sub-chiefs along the way.

Fijian society is governed by a complex system of protocols and courtesies that have been handed down through the centuries. It revolves around a Chiefly system which acts as a parallel form of government to that installed by the colonial Europeans. It is just as important to observe this correctly and in many cases it wields the greater power. To visit any areas off the beaten track without paying respectful calls on the relevant chiefs is to incur offence and a withdrawal of co-operation. Common courtesy entails the all-important *sevusevu*, a greeting ceremony in which the roots of the pepper plant *Piper methysticum* are a vital gift. To arrive at a chief's *bure* without them is infinitely worse than forgetting a box of chocolates for a hostess with whom you are staying the weekend. The roots are known as *waka* and contain a mild narcotic which is the active ingredient in the elaborate *kava* drinking ceremony that forms the central part of the full *sevusevu*.

Armed with my letters and a large bundle of *waka* wrapped in pages from the *Fiji Times*, I embarked at four o'clock the following morning on the ancient bus bound for Natovi on the north-eastern coast of Viti Levu. The warm climate makes windows unnecessary and it is easier to wave at passers-by and heave luggage in without them, so the bus had none. I was accompanied by Kelera, the delightful Fijian fiancée of an

Island Enigmas

On the Pacific's isolated archipelagos many unique creatures have evolved.

1 **Giant prehensile-tailed skink**: At just under three-quarters of a metre, *Corucia* is one of the largest of all skinks. Found only in the Solomons, it uses its tail to clamber about strangler fig trees in search of the succulent leaves of the *Epipremnum* vine.

2 **Giant lobelia** (*Cyanea*): Art Madieros stands next to a Hawaiian lobelia, whose tiny ancestors have evolved into trees.

3 **Anianiau** (*Hemignathus parvus*): The delightfully named little honey creeper is found only on Kauai in the Hawaiian Islands.

4 **Solomon giant frog** (*Discodeles guppyi*): The juvenile on its parent's head is larger than most adult frogs in temperate regions.

5 **The collared lory** or *kula* (*Phigys solitarius*): Restricted to the Fijian islands, this bird feeds on nectar and pollen from flowers.

6 **Satin flycatcher** or silktail: A velvety-feathered ornithological puzzle common only on Taveuni Island in Fiji, *Lamprolia victoriae* appears to be unrelated to any other bird.

7 **Carnivorous caterpillar**: A new discovery on Hawaii. This bristly liverwort grappler is the only caterpillar in the world to snatch and eat flies out of the air.

8 **Fruit bat** (*Pteropus tonganus*): This white-faced variety of the Tongan flying fox appears to be restricted to Fiji.

9 **Marsupial cuscus** (*Phalanger kraemeni*): This species, recently discovered in the forests on Manus, is endemic to the Admiralty Islands.

10 **Tree kangaroo** (*Dendrolagus*): Seven species occur in northern Australia and New Guinea. They were unable to cross water gaps and colonize distant islands, though humans may have taken them to New Ireland.

11 **Cassowary** (*Casuarius casuarius*): These huge birds were unable to fly or float to islands and so are restricted to New Guinea and northern Australia. Here a male guards the nest.

12 **Blue-tailed skink** (*Emoia*): In contrast, lizards like this one are found throughout the Pacific. Their small size enabled them to drift on wood to distant shorelines.

13 **Gecko** (*Gekko vitanus*): These lizards also made good colonizers. House geckos may have travelled with early Polynesians as stowaways among the crop plants in their canoes.

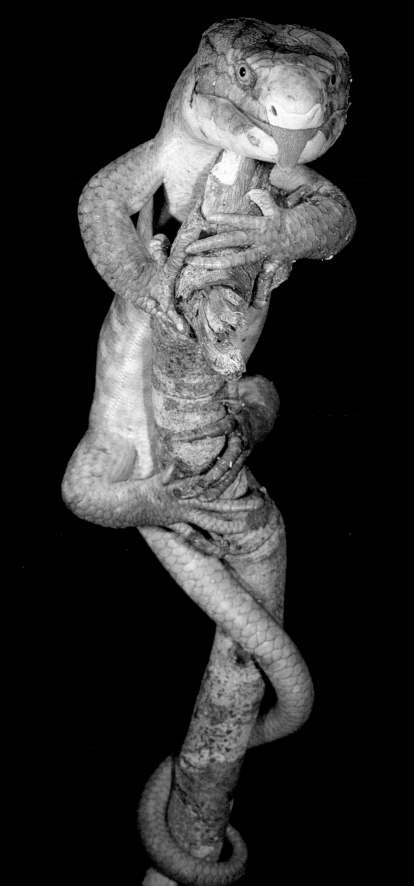

4

5

6

10

11

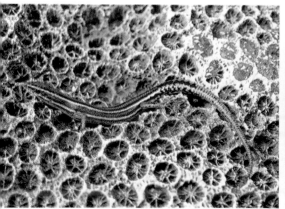

12

13

ornithological friend of mine in Suva, and Filimone from the
National Trust was to join us as guide. He was nowhere to be
seen. The bus left at five, winding through the hilly streets of
Suva as Fijians in colourful sarongs and shirts wandered down
the streets to work. A taxi suddenly careered out of a side street
and began to chase the bus, hooting insistently. The driver
ignored it; the taxi overtook and swerved in front of us, jam-
ming on its brakes. Filly climbed aboard, waved cheerily and
sat down beside us. 'Too much grog,' he whispered.

We paused at Korovou for sausages and chips and then moved
on as shutters rose on Indian stores and the first rays of the sun
caught aluminium roofs. A few hours later we arrived at the
Jubilee Ferry which would take us across to Nabouwalu on
Vanua Levu.

The sea was calm, and corals brightened the depths as we
approached. On the jetty we were met by the Buli Ravi Ravi,
the equivalent of a duke, who held jurisdiction over the southern part of the Vanua Levu. He owned the Yadua islands to
which I wished to go. In the village, immaculate in his tropical
shirt and dark blue *sulu* indicating his chiefly status, he welcomed us to his impressive thatched *bure*.

Filly composed himself to act as spokesman as we sat cross-
legged on the rush mats. He explained the purpose of our visit,
clapping four times when he had finished, and pushed over the
waka roots. The Buli Ravi Ravi clapped in return, made a brief
speech and accepted them. The simple *sevusevu* was over. A
lunch of curried chicken, fish and *dalo* followed while the
Chief's daughters fanned the food and poured water. Photographs of his family and past group events, including one
entitled *Victoria Old Boys*, gazed down at us from the wooden
beams above our heads. After lunch, the Buli Ravi Ravi showed
us the ancient Chiefly burial site above the village, explaining
that great men were often buried with their war canoes. Then it
was necessary to find an Indian with a truck to drive us to Bua,
some thirty kilometres up the coast, to secure water transport
to Yadua.

To arrive on a Pacific island and find the shops managed by
Indians, the streets full of colourful saris, and the airwaves
filled with wailing Indian music is a profound shock to many
who visit Fiji. I was vaguely expecting some Indians but I was

unprepared for such numbers or for the dominant role they play in commerce. Once I had got over my initial surprise I felt strangely at home. In my native London the cheery Indian shopkeeper and his traditionally attired wife are familiar faces of British society. Until very recently race relations here were held up as an example to the rest of the world. Most native Fijians are content to pursue an easy life. Self-aggrandizement is frowned upon; it is what you can do for your family and clan that matters. Now, however, under the new Government, Indians were wielding much greater power, and indigenous Fijians didn't like it. In Suva, disquiet over the result of the latest elections was mounting: there was marching in the streets and an Indian store had been attacked.

Like the Solomon Islanders, Fijians have the dark skins and frizzy hair that characterize Melanesians, though their features reflect their contacts with the Polynesians during the latter's migrations. The coastal plains and undulating valleys carved from the hillsides of the high islands by numerous rivers have provided flat land for the Fijians to grow coconut groves and taro, known as *dalo*. Today there are a little short of 700,000 people living here, though fewer than half of them are native Fijians. The majority are the legacy of a colonial past over which the native Fijian population had no control and a labour policy which has totally altered its future.

Fiji became a British colony in 1874. The first resident Governor, Sir Arthur Gordon, brought stable government and ruled that Fijian land could not be bought, only leased, so securing the rights of the native peoples to their heritage. The colonialists also brought measles. Most of the native population died, reducing it from 200,000 to 19,000 by 1919. The shortage of labour to work new sugar plantations meant that it had to be imported. The first batch arrived from India in 1879 to work for a shilling a day. Now Indians outnumber the native Fijians.

Whether out of a desire to impress or an insufficient comprehension of the ease with which one may enter the next world, all the Indians I travelled with in small trucks in Fiji drove at a demonic speed. Despite the fact that the yellow Toyota was now leaving a plume of dust on the gravelled track behind it that would have done justice to a tornado, the Chief sat impassively in the front as though this was a regular Sunday afternoon drive. I and the others clung on desperately in the open back. Stones leapt from the tyres at fleeing passers-by.

We crossed from Bua in the National Trust's fibreglass canoe to the island of Yadua, which stands a few hundred metres above the blue sea. The usual crowd of children and villagers rushed to the water's edge and helped us through the corals. No *papalagi*, the Fijian word for white man, had been there for a year and a half, and then it had been Kelera's betrothed, Dick Watling, who had come to survey the iguana population. The village was a perfect picture of South Sea Island tradition, now slowly fading. Almost all the huts were of palm thatch, some old and dark brown, others straw-yellow, newly built. A square patch in the middle served as a football pitch and dancing ground. Fruit trees, coconut and banana palms skirted the village, blending into the scattered remains of forest that clothed the hills behind. The scene could not have been very different from that which had greeted the first European explorers two hundred years before.

The first European to land on Fijian shores was Captain Cook, who anchored off Vatoa in the Southern Lau Group in 1779. Ten years later, Captain Bligh was chased through the islands by hostile natives in canoes on his voyage to Timor in Indonesia following the mutiny on the *Bounty*. The area through which he passed in the Yasawa Islands is still called Bligh Water. The American Exploring Expedition of 1840, led by Lt Charles Wilkes, which did much to record the early history and natural history of the Pacific Islands, created the first reliable map of Fiji. Early reports of the hostility of the natives and their apparently insatiable interest in cannibalism kept most explorers at bay, as Wilkes reported:

> The eating of human flesh is not confined to cases of sacrifice for religious purposes, but is practised from habit and taste . . .
>
> The cannibal propensity is not limited to enemies or persons of a different tribe, but they will banquet on the flesh of their dearest friends, and it is even related that in times of scarcity, families will make an exchange of children for this horrid purpose.
>
> The flesh of women is preferred to that of men, and they

consider the flesh of the upper arm above the elbow, and of the
thigh, as the choicest parts.

Human flesh was known as long pig and it was common
practice to bind the hands and feet of as many as twenty men at
a time before casting them into the *mumu* ovens alive. Once
cooked these were eaten using large four-pronged wooden
forks, with *malawaci*, a vegetable dish made from the leaves of
the shrub *Streblus anthropophagorum*, and a sauce made from
fruits rather like tomatoes which grew round temple sites. One
of the last Europeans to be eaten was the Reverend Baker in
1867. The wooden fork used for the occasion sits in the Fiji
Museum. Rumour has it that they boiled his hide boots for
hours, imagining they were his feet, before deciding that they
were too tough.

Eventually the lure of sandalwood, bêches-de-mer, and
finally potential religious converts, stirred Europeans to over-
come their fears and secure their grip on Fiji. A selection of
colourful escaped convicts and beachcombers instructed the
Fijian chiefs, who appeared constantly to be at war, in the use of
firearms, and so played mayhem with the ancient balance of
power. One chief, Cakobau, eventually dominated the rest
with the help of an adventurer named Charlie Savage, whose
skill in weaponry enabled Cakobau to subjugate most of his
rivals from his stronghold on Bau Island. Once Savage's useful-
ness was over and his excesses with women and drink had
become a bore, he was summarily clubbed to death and eaten.

Cakobau's rule was shaky and threatened by the Tongan
Chief of Lau, Ma'afu, as well as by aggressive posturing from
America instigated by its Consul, whose house had been burnt
down during a night of revelry and then looted by the natives.
In 1874 King Cakobau ceded the islands to Britain, and his war
club which had dislodged so many brains was sent to Bucking-
ham Palace as a gift to Queen Victoria. She returned it decorated in silver and gems and it stands to this day as the symbol of
power in the Fijian Parliament.

As I leapt out of the canoe on to Yadua's sandy beach, a tall
bullet-headed gentleman snapped to attention and shot out a
hand. Filly informed me that this was Anare, the warden of the
sanctuary on the nearby island of Yaduataba where most of the

iguanas were still to be found. After a brief break for tea and biscuits at Anare's house, I was ushered to a large square hut to be presented to the Chief. Here I found a full meeting of elders in progress. The Chief sat in T-shirt and *sulu* at one end, his speaker on his left, the *matani vanua* or master of ceremonies on his right. Above his head large beams straddled the roof, one bearing the usual photographs and framed proclamations. On the rush mats opposite the Chief was the *tanoa*, a huge round bowl on four legs carved from a single piece of *vesi* wood. Two servers sat on either side of it and behind sat the man whose job it was to stir the brown fluid in the *tanoa* and examine it for strength and consistency. Behind him about thirty men sat cross-legged, listening to the speeches from the other end. From the back of the room, a crowbar thudded into a solid wooden mortar to a steady rhythm, grinding *yaggona* roots to dust. The pounded fibres would be put inside a sackcloth bag into a *tanoa* filled with fresh water. Kneading the bag leaves the fibres but flushes out the powder, turning the water a muddy brown. Anare, Filly and I took up a place in the crowd to one side and waited our turn to speak.

The *matani vanua* soon announced us, and Anare, passing over the *yaggona* roots I had bought at the market place in Nabouwalu, explained the purpose of my visit and requested the Chief's approval. At the end there was a series of long drawn-out 'Whoooohs!' and 'Aaahhs!' in unison from the men, followed by methodical handclaps. Permission was granted and the business of the morning continued. This concerned the payment of dues to refurbish the small church. Five dollars were required per man; some had not yet paid up. I offered my contribution and there were cries of 'Vinaka, vinaka'. My name was duly entered into the ledger. The *matani vanua* called 'talo', and the server wiped the rim of the *tanoa* three times before filling a huge half-coconut known as a *bilo* with brown liquid. It was passed to a second server and then on to me. After three respectful claps I took it in both hands and drank it down in one, as is the custom. The *kava* tasted bitter and the inside of my mouth went numb.

The ceremony continued for several hours, during which I heard translations of stories concerning the iguanas with an increasingly light head. As these sessions can go on until dawn the following day, the Chief must have the constitution of an ox, though Filly informed me that he used a few tricks to prevent himself from passing out. Hungry for legends I asked if there were any concerning the iguanas. Apparently there were not. Like the spirits they had always been there, though they were to be feared and might stick to the skin if they leapt on to an unsuspecting villager in the forest. There was greater interest in the fact that the village was not receiving a cut of the money paid by the Trust to a higher area Chief for the reserve's upkeep. Goats had once roamed the island of Yaduataba, destroying the vegetation on which the iguanas depended, and had had to be forcibly removed when the reserve was created, not without a little disgruntlement on Yadua.

After a lunch with Anare of fish and taro leaves in coconut cream, we motored on to Yaduataba in perfect sunshine. Passing through the narrow gap between the islands we floated over magnificent reefs of coral, reflecting blues and greens and splashed with colourful fish. On the eastern edge of the main island tall forest grew, and a few spirals of smoke rose from a collection of grass huts on the beach. Rounding the northern point of the iguana island we came to a beautiful sandy beach on the western side and landed there before going inland to explore the forest. It was not difficult to find the iguanas, which were lolling on the branches of hibiscus trees a few hundred metres from the shore. They eyed us lazily from their lofty perches, occasionally moving behind larger branches to avoid being seen, their emerald-green bodies and lighter bands a perfect camouflage once they climbed higher into the foliage. Today there may be as many as 6,000 iguanas left on the island.

I climbed a trunk and began to edge out along a branch towards a somewhat lazy specimen which appeared uninterested in escape. Just as I thought the branch, which was quite a distance above the ground, would break and reached across to capture the placid beast, I heard a 'hsssssh!' It was Kelera below. 'There's one of those horrid nests below you. Be careful.' I looked down and saw a ball the size of a coconut with a number of angry hornets buzzing around it, an arm's length from my rather baggy shorts. Having got this far I was not to be put off and gingerly grabbed my prize, which instantly came to life and clung to the branch with unexpected strength. A tug-of-war ensued in which I began to sway up and down in front of the hornets with a less than mesmerizing effect. The iguana had turned black with rage. Finally I yanked it free and almost fell to the ground, thankfully unpursued.

John had told me that the eggs of this species have the longest incubation period of any iguana in the world. The discovery was made by Ivy Watkins, a remarkable English woman who looks after the animals at the Orchid Island Cultural Centre outside Suva, begun by her husband. A number of the iguanas roam at will around her house and one of these had once laid eggs. She placed them in a plastic tub of sand in a warm cupboard under the sink in the bathroom. Eight months passed before she opened the door to discover the cupboard crawling with baby iguanas.

Ivy Watkins' bathroom cupboard remains the finest rearing unit in the world for crested iguanas. Tarronga Zoo in Sydney, one of only two zoos so far to breed them, is unable to match her success in spite of all its high-tech incubation equipment.

A light breeze was sending a shiver through the palm fronds above my head. It was not difficult to imagine the life of a castaway here. Behind me Kelera was poking the fire from which a delicious smell of baking mud-crabs, collected earlier from a mangrove creek nearby, spiralled into the night. A simple tarpaulin would provide shelter under the sky. Filly was casting a fishing line into the shallows: I could see white splashes in the warm darkness. On the left-hand side of the bay, the hillside carved a line across the stars. Out to sea the Pacific swells roared, leaving a dim white line of surf. I tried to imagine the scene that must have been played millions of years before.

Six thousand miles away on a West Indian or perhaps a South American coast, the last of the hurricane season's storms wrenches the canopy of a mangrove forest. Rain rushes in sheets across the glistening landscape. A brown and swirling river cuts into the bank and with a crash a tall mangrove tree topples into the swollen stream. Carried down with a tangle of others, the tree turns in the current. Several adult iguanas are crowded on to it, along with some tree mice, smaller lizards, spiders and insects. Days later, locked into a raft of vegetation, it drifts far from land, carried on the cold Humboldt current northwards to meet the Southern Equatorial current moving slowly west. One by one the castaways die. The mice are first to go, being unable to stand the sea, and once the insects and spiders have been eaten the smaller lizards slip from the branches. Only the iguanas continue to feed on the mangrove leaves which, used to the salty environment in which they grow, live longer here than could those of other trees. Weeks pass before the leaves are used up and the iguanas weaken. In desperation a pregnant female lays her clutch of eggs inside a hollow branch high above the water before she too dies. After months of voyaging the current carries the tree north of where Tahiti and Samoa are today and then turns south. Bleached by the sun, this tree floats on a calm day into a sheltered lagoon to lodge in the sand. Weeks or even months later, baby iguanas emerge and scamper ashore, disappearing into the forest in search of food.

The red mangrove tree *Rhizophora mangle* too appears to have floated from the Atlantic side of South America where it originated, through the Central American gap before it closed 5,700,000 years ago, to reach as far west as Samoa.

The following morning I took some of the iguanas out into the lagoon and placed them on a large floating log to see how they would fare. Some instantly leapt into the sea and swam expertly under the surface towards land; others could swim adequately across the surface, their limbs held close to their bodies, tails thrashing the sea. Returned to the log after a few such duckings they clung to their raft like prehistoric monsters, but it was not long before their movements slowed and their grip weakened as the wind chilled their bodies and the sea splashed over them. Hurriedly I carried them to the reviving warmth of the dry sand near their trees.

As we left the bay I wondered at the achievement of these Pacific castaways that had populated the islands, now so simple to reach by plane and canoe. I could not help thinking that the rafting theories were too simple. The current flow was wrong; it was all too fantastic. Not one person I met had ever seen a large floating mat of vegetation far out to sea with even small skinks surviving on it. Most of them would break up anyway shortly after they left the coast. It also seemed that the adults would not last long if constantly drenched by waves at sea. Could the iguanas have walked across the Antarctic from

South America before it broke free from Australia and then on to Fiji? But Fiji was never joined to Australia, and neither these iguanas nor their fossils are found in Australia now.

Elsewhere evolution was taking a different course. In Suva news was breaking of the first petrol bombs beginning to fly as Fijian democracy burst at the seams, and the chants of the *Taukei* movement split the night: 'Fiji for the Fijians!' As our new canoe cut across the smooth surface of the lagoon, the ripples began to spread outwards, away from the island where castaways from a natural world had found security, and out into a turbulent sea.

Four days after I left Fiji, Colonel Sitiveni Rabuka and his troops entered the Fijian Parliament and removed its newly elected Government at gunpoint. The coup was almost blood-less. Western developed nations were outraged: the colonially instilled notion of democracy had been overthrown. Fiji, the perfect example of harmonious racial co-existence for all to follow, had failed. Leaders of native peoples across the Pacific gave a knowing smile: an example indeed.

Chapter Six　　**TO ISLANDS ON WINGS AND WIND**

Taveuni and Samoa

> You lie on a mat in a cool Samoan hut, and look
> out on the white sand under the high palms,
> and a gentle sea, and the black line of the reef a
> mile out, and moonlight over everything . . . It
> is sheer beauty, so pure that it is difficult to
> breathe in.
>
> Rupert Brooke, *Collected Poems*, 1918

here is often an uneasy calm preceding a hurricane's touch. Light airs are scooped from warm seas into the upper atmosphere, and pressure near the surface drops like a stone. Beyond the horizon, the sky darkens to a bruise. Warm, moist winds rush in to fill the rising vacuum, the anvil-tops of a hundred storms join arms, and the air of the whole region begins to turn, slowly at first, then gathering momentum and strength until a vortex is born, writhing at its centre and producing sucking winds far stronger than any gale. Like a slowly revolving Medusa the cyclone advances, flicking the sea into a foam, churning the surface of the Pacific swells into a carpet of twisting spume and the waves themselves into cliffs. Latent heat oozes from smoke-grey clouds as they rise, cool, and condense into sheets of rain. Soon the first tendrils of the hurricane begin to cast a shadow across the green peaks of an island in its path. Freshening breezes begin to tumble through the sturdy rainforest trees, enveloping the palms swaying urgently above the sand.

With cracks like pistol shots the palms begin to snap, and as the wind turns to the hillside, the buttress roots of mightier trees are torn apart; tall crowns topple to the sound of thunder. Lightning moves across a scene of growing devastation. Whole hillsides are laid waste. The air is filled with a fleece of leaves, flowers and fruits, stripped from a million branches. Unable to cling to the buffeting forest, a flycatcher foolishly takes to the

air. Its tiny wings are no match for the cyclone's energy and it is
swept far out to sea. Over the mountains the wind collects an
ever-increasing detritus of small insects, spiders, organisms
too small to see; the dust of life is swept from stream banks,
branches and the seashore. Pastel-pink pigeons and bright-
green lorikeets are swept hopelessly upwards into the vortex
and joined by clumsy bats, tumbled from their daytime roosts
in their hundreds unable to cling to the branches, some carrying young, some yet to give birth. All try uselessly to fly
against the force which carries them from the land.

When the wind has moved on and the branches stand bare,
birds and bats alike will begin to starve in the forest, and fall to
struggle on the ground. Weeks will pass before the trees grow
fruits again. Meanwhile, far out to sea, sucked into the atmosphere and swept on the uncaring hand of the wind, are the seeds
of new life, the spawn of a dying forest, the colonizers of new
islands.

Exactly where most birds have come from that brighten the
forests of the remote Pacific Islands is a question that keeps
biogeographers arguing long into the night. With the exception
of a few species in Hawaii and those on the eastern Pacific
Island groups of Juan Fernandez, Easter Island and Galapagos,
none of the land birds of the South Pacific is of American origin.
There are, of course, many sea birds which wander freely
throughout the world's tropical and subtropical seas, and in
addition migratory species which pass through the islands on
their way to or from their feeding grounds. Of these the now
extinct Tahiti sandpiper *Prosobonia leucoptera*, found by Captain Cook but not seen again since, and a similar species which
still exists on the numerous atolls of the Tuamotus, may have
colonized the islands from the Arctic. Turnstones, bristle-
thighed curlews, and wandering tattlers are still to be seen
probing the muds and sands of the South Pacific each year,
having flown south from their Arctic breeding grounds. It is in
the beautiful Palauan Islands just east of the Philippines and in
the Mariana Islands that the greatest Asian influence is to be
found. The yellow bittern *Ixobrychus sinensis* is prominent in
the marshes while the jungle nightjar may have island-hopped
down the Mariana chain to Palau from northern China or
Japan. Reed warblers also took this northern route through

Micronesia and now eight species and many subspecies are found scattered across the ocean as far as the eastern Pacific. Swiftlets, woodswallows, short-eared owls and several rails may have entered Micronesia from the Philippines and Indonesia, though some could have an Australian origin.

Most of the Pacific's ancestral birds journeyed from New Guinea. Even the most distant islands such as Pitcairn and Hawaii have birds which came from here. Their colonizing route across the Pacific seems to have taken them from New Guinea to the Bismarck Archipelago, on to the Solomons, Vanuatu and New Caledonia, to Fiji and Samoa, east to the Society Islands, and lastly north to the Tuamotus and Marquesas. The megapodes, fruit pigeons, kingfishers, white-eyes, weaver finches, cuckoo shrikes and honey-eaters all came this way. As usual their numbers dwindled as they flew eastwards; in Tahiti there are just twelve land birds while an island of similar size in the Solomons or Vanuatu might support forty. The birds of New Zealand, New Caledonia and Vanuatu appear to be of Australian origin, as do some of those in the Solomons and Fiji. Only in New Zealand, Hawaii and the Galapagos Islands are there endemic families of birds, indicating their long separation. Almost all other Pacific birds belong to genera related to those of New Guinea or found there.

A number of birds in the Pacific are flightless, yet their ancestors probably flew there. The rails, inconspicuous brown birds with long beaks and longer legs that carry their out-stretched bodies rapidly across roads to vanish silently into the undergrowth, can barely fly. The cormorant of the Galapagos Islands holds out wings to dry in the sun which are pathetically small compared to those of its fish-eating relatives elsewhere in the world. To fly requires such great effort that once wings are no longer needed to forage for food or to escape predators they are, in evolutionary terms, quickly dispensed with.

In the vaults of the most prestigious museum in the world, the British Museum of Natural History, there is the skin of a bird named after a Scotsman. It was collected over 130 years ago on the island of Gau by a Dr F. M. Rayner on an expedition to the Lomaviti Islands in the Fiji group aboard HMS *Herald*. It was sent in a box to London where it was named MacGillivray's petrel after the ship's naturalist who, unknown to the closeted museum taxonomist who described it, had left the expedition under a cloud. The bird is related to the albatross, but possesses a dark plumage and probably nests on cliff tops; since the day it was discovered it has never been collected again. The name has persisted but the bird has not – or has it?

A seabird restricted to a single island is very unusual; most travel widely in the Pacific. The specimen in the British Museum, a juvenile, showed that the birds may have bred in the month of October, when the fledgling was collected. Petrels tend to nest on inaccessible cliff tops emerging from the forest in the centre of remote islands. Gau was just such a place, and in October 1983 ornithologist Dick Watling set out to try to find the elusive bird. Petrels tend only to fly at night, and after an exhausting march into the mountains in the centre of Gau, the search began by the light of a powerful lamp. After hours of fruitless scrambling over branches on the treacherous cliff tops, Watling was amazed when a large bird suddenly flew into the torch beam, collided with his head, and fell stunned to the ground. After 128 years, MacGillivray's petrel had been found. The story illustrates the ridiculous state of our knowledge about the rapidly disappearing birdlife of the Pacific.

Dick and I sat on the rush mats spread over the floor of his house in Suva, the capital of Fiji. I had known him since we had tramped round the rainforests of Sulawesi in Indonesia together in 1981. Now he had returned to Fiji, where he had been brought up, and earned a living as an environmental consultant.

'The landbirds in this part of the world are some of the most beautiful I've seen, yet there's no one studying them, and I can't cover them all. I'm not a millionaire. More grog?' He offered me a coconut full of *kava* which I downed with customary claps.

'The islands are too remote and complicated to reach and half the species will be extinct due to land clearance and the mongoose before we know anything about them.' The mongoose had been imported to Fiji from India to kill rats but turned on the native birds instead. Eight species have been extinguished on Viti Levu and Vanua Levu alone.

The pigeons of the Pacific are some of the most spectacular in the world. They have enchanting names such as the many-coloured fruit dove, which looks as though an artist has thrown colours at it with his palette knife, the crimson-crowned fruit dove, the golden dove and the friendly ground dove. All but the last belong to the *Ptilinopus* genus of pigeons which have evolved into species with arresting bright colours and unusual voices, and are restricted to just a few islands in Fiji, Tonga or Samoa. Of their nesting habits, diet or numbers, almost nothing is known. By far the most superb of these pigeons was the dove reputed to fly like a flickering flame. I was determined to see one.

'The best place to search for them is Taveuni,' Dick told me. 'The mongoose hasn't reached that island so most of the original species are still around. It's called the flame dove because it looks like a flickering flame as it flies the forest. The silktail's also there. Nobody knows what that is; it looks a bit like a small bird of paradise. If you see anything with a white bum flying about low down in the forest, that'll be it.'

Taveuni is known as the garden island of Fiji. It is 16 kilometres long and the young volcanic hills in the interior rise to 1,241 metres at the summit of Mount Uluiggalau. Volcanism ceased on the island less than 2,000 years ago, but since then it has been colonized by a rich flora. Abundant rainfall produces some spectacular waterfalls in its forests, the branches of which are festooned with ferns and orchids as well as the indigenous *Medinilla spectabilis* which hangs in clusters of small red bells from the trees. High in the centre of the island is Lake Tagimaucia which is famous for its red and white *tagimaucia* flowers (*M. waterhousei*), growing by the lakeside. Legend has it that the flowers, which resemble tears in the morning dew, are those of a girl whose irritable father wished her to marry an old man whom she did not love. She fled to the lake for solitude and her tears turned into glistening flowers. When her father found her and saw the delicate flowers, his heart was softened, and the girl was reunited with her lover.

A circular rainbow pursued us through a patch of leaden clouds above the Koro Sea as I flew towards the islands. Near the coast lies the Chiefly Island of Bau, the inhabitants of which are still accorded almost royal status. Further north the large volcanic island of Ovalau slipped under our port wing. On its east coast is Levuka, a former whaling settlement and the colonial capital, where the agreement for the annexation of the Fijian islands by Great Britain was signed. To the south lay Gau island, a distant shadow in the mist, and then below us the island of Koro came into view, surrounded by the aquamarine of its coral reefs. Forty minutes more passed before the Twin Otter, decked out in the royal blue and white colours of Fiji Air, swooped into Matei airstrip and bounced on to the grass.

In the small hut which served as the airport lounge, Indian taxi drivers jostled for business. The one I chose drove with customary terrifying speed down the beachside track through the mangrove forests to Somosomo, capital of the island and home of Sir Ratu Ganilau, Paramount Chief and the Governor General of all Fiji. I purchased some *waka* and descended on Kelera's father, Elio, a seventy-year-old veteran of the Solomon Islands war, which had cost him an eye. He kindly agreed to find me a guide and the following day I set off up the mountain accompanied by the tall half-Tongan, half-Gujerati 'Jim Boy' to search for the flame dove.

Jim Boy spoke excellent English but had a somewhat melancholy air. He had travelled widely before settling as an Indonesian chef at the Java Restaurant in Suva. He surprised me by indicating that he now had eleven wives – which he later modified to eleven women and a child from each of them. Now he lived on Taveuni and had only a taro patch, but time to show people around.

As we climbed up through hillsides cleared of trees by a large bulldozer operated by a nameless German, we passed collections of young coffee plants waiting to replace them. A Fiji goshawk settled on a stump before wheeling into the sunshine, it's 'weeeee-weeee' calls floating on the cool morning breeze. This elegant bird is a frequent sight in open wooded country, but only in Fiji; it is found nowhere else. Its slaty wings outstretched, it flew with a combination of wing beats and glides across the clearing, its salmon-pink breast contrasting with a grey head and yellow eye. Near the forest edge a male Vanikoro broadbill suddenly sped outwards, gunmetal-blue head prominent on dark-blue wings and rufous orange breast. It seemed the goshawk had seen it and turned as if to strike but the broadbill flew up at it and fearlessly attacked. The noble goshawk wheeled and turned, and eventually flew away.

The Vanikoro broadbill is one of the few birds to have adapted well to the breaking up of Fiji's forests and is commonly seen in town gardens and suburbs. It is a form of flycatcher which flits through the forest picking caterpillars and beetles from the undersides of leaves, hovering before them as it does so. It is an artistic nest-builder, creating a small cup of grass stems and fine fibres decorated with lichens, pieces of moss and the odd leaf. The whole thing is bound with silk from spiders' webs. Horse and cattle hair furnish the lining into which two speckled white eggs will be laid. The bird's only relative is found on the Santa Cruz Islands in the Solomons. Within Fiji it has a number of subspecies on different islands, all varying slightly in size or colour. Even though the broadbill must somehow have reached Fiji from lands further west, the short distances between islands in the Fiji group have been sufficient to isolate each population. This kind of reproductive isolation – even over comparatively short distances – can often lead to the creation of completely new species which, though all related to the ancestral stock, may appear quite different in shape or colour. This is the kind of process which gave rise to Darwin's varied finches in the Galapagos, and which has resulted in Fiji's magnificent *Ptilinopus* pigeons of which the flame dove is but one.

It still eluded me. We climbed higher, at last reaching the line of undamaged forest, entering the cool shade with relief and beginning the scramble over tree roots, pushing aside tree ferns and tangled vines as we pushed up the small track towards Lake Tagimaucia. On the way I narrowly missed brushing against the large leaves of the *salato*, one of the few poisonous plants in the Pacific Islands. This is like a nettle the size of a tree, with fine hairs over its leaves which contain poison. If injected, this causes frightening welts and intense pain which recurs for weeks. Fortunately the islands of the South Pacific have few such plants as well as few snakes and biting insects.

Rounding a bend in the track, we came to a clearing filled with squawking and guttural sounds which could only come from the Taveuni parrot. A number of them were walking backwards and forwards along the branches of a fig tree, while another pair were demonstrating their abilities as trapeze artists on hanging vines. The musk parrots of the Fijian Islands are quite simply the most handsome parrots in the world. *Prosopeia tabuensis taviunensis* is confined to the islands of Taveuni and Quamea alone, and four of them were cavorting in front of me, quite indifferent to the fact that I was there. Their backs and wings were an iridescent emerald green rimmed in sky blue. The undersides of their tails were black, while the heads and breasts were the deepest maroon. As they gathered at the base of a large bough to watch, bobbing and chortling, they looked like a collection of distinguished gentlemen in their smoking jackets swapping jokes at the club. Another species found only on Kadavu has a brilliant red head and breast with a sky-blue collar, a feature distinguishing it from other parrots on Vanua Levu and elsewhere. Formerly, the sulphur-breasted musk parrot *P. personata*, which is similar but with a black face and bright-yellow breast feathers, was the only large parrot on Vanua Levu, but the red-breasted species was introduced there many years ago.

Sometimes these birds will gather in feeding flocks of forty or more to reach for mangoes, guavas or fruits of the *ivi* tree. Unlike pigeons, parrots are generally predators of seeds, tending to crack them in their powerful bills to reach the kernels and so destroy their chances of germination. Only the seeds of figs and other such soft and small-seeded fruits are likely to be consumed and then distributed around the forest in droppings. Fig trees are an important source of food for many forest birds in the Pacific. They grow to an enormous size, though the fig tree starts out as a tiny epiphytic plant sprouting in the branches of a host tree into which its seed may have been dropped by a passing bird or bat. Its roots descend to the ground and, once there, swell with the moisture and nutrients the soil provides. Eventually these will completely encase the host tree and strangle it to death. The banyan fig eventually stands on the curtains of roots it has sent to the ground. Fortunately these trees are hard to cut down, and are consequently left alone by commercial loggers and farmers alike, providing food for the diminished population of remaining birds.

Ahead the path steepened through the tall trees, and I began to feel the weight of the small pack on my back. It was now well into the morning and Jim Boy beckoned me to quicken my pace. My vision was soon obscured by biting trickles of sweat, and I could feel the first coatings of damp moss on vines and branches as we made our way higher up the mountain's flanks. Numerous small birds tantalized us with their calls but refused to be identified, merely offering a flash of yellow or iridescent blue as they disappeared through the undergrowth. Worst of all, I could not hear the 'tock-tock-tock', like the dripping of a tap, which would announce that a flame dove was nearby. A casserole-sized bird took off in a blur of wings from a tree crown: a Pacific pigeon. It settled in a tree below us, the distinctive blue-black knob prominent on its beak. It looked jerkily in our direction for further signs of danger, with good reason. Jim Boy explained that the pigeons were attracted by smoke. A small fire beneath a tree would bring in three or four which he could readily despatch for the pot with his .22 rifle. I wondered that there were any left.

In Fiji the barking or Peale's pigeon is confined to the larger islands alone, while the equally large Pacific pigeon tends to be found on the smaller islands. The latter is known as a 'tramp' species because of its habit of wandering between islands. It is a non-specialist, finding food where it can from a variety of trees and vines. This ability has enabled it to colonize the tropical pacific from the Bismarcks east to the Cook Islands. In Micronesia another large pigeon with a handsome green back, white head, and rust-coloured underparts reigns supreme. Tahiti and the Marquesas each have their own endemic species of these large, heavy pigeons as well. *Ducula goliath*, an enormous chestnut-brown pigeon found only on the island of New Caledonia, must be the largest I have seen; it would make a satisfying meal for any aspiring David.

Savai'i and Upolu in Western Samoa are the only places in the world where you will find the tooth-billed pigeon. It is rare, extremely hard to find and must be in danger of disappearing from the world. When Ramsey Peale first described the curious pigeon he found in Samoa in 1848, he was so struck by its beak, which resembled that of the dodo, that he named it *Didunculus* or 'little dodo'. Since then others have believed the bird to be related to parrots because of its eating habits and apparent ability to hold food in its feet. In fact, this is not so. The tooth-billed pigeon has evolved in a similar way to parrots, but it is still a pigeon. Only its beak resembles that of a dodo;

nothing else. The large orange beak and chestnut back serve to distinguish it from all other pigeons on Samoa. It feeds almost exclusively on the fruits of the *Dysoxylum* tree. These tall trees produce clusters of round green fruits a little smaller than ping-pong balls. Grasping the fruit with the upper half of its beak, the pigeon moves the lower half backwards and forwards like a saw. It is able to cut even hard, unripe fruit, to get at the four seeds inside. Once the fruit is open, the highly mobile top mandible manoeuvres the seeds into the bird's mouth.

Almost certainly the survival of the *Dysoxylum* tree is as intimately connected to this pigeon as that of the *Calvaria* tree was to the dodo. For almost three centuries following the dodo's extinction no *Calvaria* trees germinated on Mauritius. No bird on the islands appeared capable of opening the enormous hard-shelled fruits so that the seeds could germinate. Scientists concluded that the dodo must have been responsible for this task in the past. The handsome *Calvaria* trees, for which the island was famous, seemed doomed. Only when seeds from the last remaining trees were fed in desperation to turkeys, which had crops strong enough to weaken the seed casings, did the first seedlings grow: the tree was thus saved. Many birds, including pigeons in the Pacific's dwindling forest, have a job to do; they do not exist simply to grace the hunter's table. Evolution has charged them with the means to distribute the offspring of trees and so ensure the survival of both tree and bird. To harm one half of such a partnership is often to threaten the survival of the other.

A secondary invasion of birds may eclipse the first colonizers by competing with them for food, reducing the original species to mere relics on isolated islands. It may be that the tooth-billed pigeon represents just such an early invader now living in reduced circumstances. Once a species has become established on an island, it might disperse to others, each of which gives rise to its own unique form. If there is no competition, the initial colonizer might evolve into many species, adapted to exploit the variety of niches on one island or several; each one will be quite different in colour, shape and form from the original colonizing ancestor. In the Pacific, all of these things have happened.

It was early afternoon before Jim Boy and I reached a cleared look-out some distance from the summit of the mountain. As the *tagimaucia* flowers did not bloom until December, there seemed little point in continuing to the crater lake where they are found. Instead I took time to rest against the blood-red leaves of young ferns which grew there. Disappointed that I had not managed to see a single flame dove from my high spot, I called to Jim Boy and we set off down the mountain. By four o'clock we were about to emerge from the forest when I heard a sound, a single penetrating 'tock'. In the forest many birds seem to have the qualities of a ventriloquist; their calls appear to come first from the left and then from the right, making them very hard to locate. I craned up into the treetops, scanning the branches with my binoculars, and suddenly I saw it. No photographs exist of this pigeon and the paintings I had seen left me quite unprepared for the brightness of its plumage. It was a fluorescent orange, which stood out against the leaves like a traffic signal. The bird was male; a slight tinge of green dusts the male's head, whereas the females are a dark olive-green all over. Soon other males joined it in trees nearby and the canopy was filled with intermittent 'tocks', each delivered at the apex of an imperial bow, as the birds displayed. I was overjoyed, but more was to come. As I squatted in the undergrowth another of Taveuni's specialities revealed itself, sporting a bright white rump in contrast with black almost fur-like feathers: a silktail, stopping on a branch not ten feet away as it paused before vanishing into the undergrowth of the only island where it has evolved.

Not all creatures that have wings are as powerful fliers as birds. The Pacific Islands are full of insects, including attractive butterflies, beetles and flies, which would certainly never have set out intentionally on long inter-island journeys. Many of them also have bats, which prefer to confine themselves to forests and caves. Some islands in the west Pacific enjoy the attentions of small insectivorous bats; these have failed to reach the islands of the Central Pacific, though one species appears to have colonized Hawaii from America. The islands of the South Pacific are the domain of some of the largest bats in the world, the Pacific flying foxes. One species has reached as far east as the Cook Islands, though it was almost certainly carried there by Tongans in their canoes. Another is mainly restricted to the islands of Samoa; this species is one of the most endangered. How the Samoan flying fox got there remains a mystery, though it is almost certain it could not have done so by using its wings alone. I went to Samoa to see if I could find out more.

Polynesians arrived here more than a thousand years BC. In AD 950, the Tongans invaded and ruled Samoa for three centuries. Six hundred years later new, more subtle invaders had arrived. The first was the Reverend John Williams aboard the *Messenger of Peace*, from the London Missionary Society. Williams brought religious influence and influenza. The Germans followed and developed commerce and copra, while the Americans, eager for new territories, imported a consul and a Constitution. In 1875 the British sent a gunboat and attempted to remove them both. Civil war then raged.

On 16 March 1889, the US warships *Trenton*, *Nipsic* and *Vandalia* faced the German *Alder* and *Eber* in Apia harbour on the north coast of Upolu in what is today Western Samoa. The British had also ordered HMS *Calliope* to enter the capital's harbour to add weight to their claim on the territory. While the Germans, Americans and British postured for power and sovereignty over a small and defenceless tropical island in the South Pacific, an even greater storm was brewing offshore. A terrible hurricane was about to strike Apia.

Despite warnings of worsening weather, none of the captains wished to leave the harbour for the safety of the open ocean lest another might take the island. Almost too late, the British captain sensed the danger. At the last minute, he ordered his ship to weigh anchor and the *Calliope* began to turn into the rising sea.

Walking along Beach Road on the Apia waterfront today there is nothing to be seen of the hurricane which buffeted the town a century ago or of the wrecks of the German and American ships which littered the harbour then. The *Trenton* and *Nipsic* were thrown on to the reefs, the *Vandalia* sank, the *Alder* turned upside down, and 200 lives were lost. Only the *Calliope* survived by bravely beating out to sea in the face of the storm. Apia is now a pretty town dominated from the sea by the twin white towers of its Catholic cathedral set in a line of

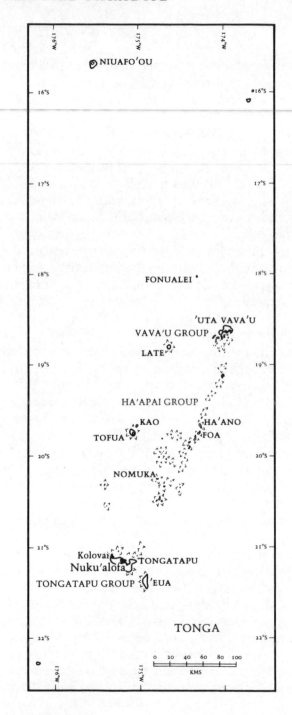

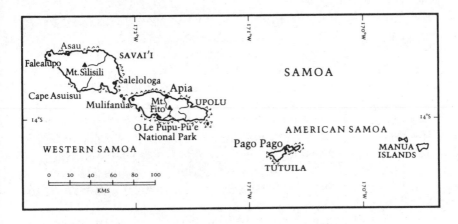

low colonial bungalows, gracious government buildings and a
few smart hotels, backed by steep green mountains rising from
the coast. More than 33,000 people live in the capital now. The
Germans finally raised their flag over the islands ten years after
the hurricane, before being pushed out by the New Zealand
Expeditionary Force at the beginning of the First World War. In
1962 Samoa was the first Polynesian nation to gain independence, and His Highness Malietoa Tanumafili II, Paramount
Chief of Samoa, has been Head of State ever since.

Tradition is all important to the *matai* or chiefs of Western
Samoa. They police the two large islands of Upolu and Savai'i
which comprise the Samoan nation and make all decisions of
high office. They alone can stand for seats in Parliament; just
two remain open to non-Samoan residents. The Fa'a Samoa, or
Samoan Way, has persisted despite a Government on the verge
of bankruptcy and dependent on aid from Japan, Australia and
New Zealand, with little to export other than copra, taro and
cocoa. Samoans are proud, tall, and honey-skinned. They have
black wavy hair and, unlike the islanders of Fiji, they are
Polynesians; the Maoris of New Zealand are the only Polynesians to outnumber them. Adherence to tribal protocol is
almost feudal, religious customs are vehemently observed,
success and wealth are attributes of the communal *aiga* or
family, and individual success outside the Chiefly system is
frowned upon.

Air travel has enabled more Samoans to exist outside the
islands than in them, principally in the US and New Zealand. A

consumer boom is fuelled by television from Pago Pago in the neighbouring American Samoan Islands. These were annexed by the United States in 1900 for Tutuila's magnificent harbour, something the US has never let go. The contrasts between American Samoa and Western Samoa are stark. Pago Pago is a miserable town where people move in fear of insanely-driven trucks, and unsmiling Samoan youths hide behind mirrored sunglasses and clutch ghetto-blasters blaring American funk. Traditional canoes have been replaced by sleek American-owned tuna catchers which assemble here from all over the Pacific. As a result Paradise stinks of putrefying fish from the processing factory across the bay.

While both groups of islands retain many traditions away from the towns, it is in Western Samoa that they are strongest, and so the conflict with modern desires is greatest here. Western Samoa is still a paradise, yet its young show one of the highest suicide rates in the world, a symptom of the trauma which taints this beautiful land. Many are educated in New Zealand, and once they have experienced the freedoms that the Western world provides, cannot come to terms with the restrictions tradition dictates in their own. Faced with years of virtual slavery to their families' will, and no honourable way out, many succumb to depression and take their own lives with Paraquat; this is the ridiculous paradox of a land occupied by what Brooke called 'the loveliest people in the world'.

It was a bright sunny day as I strode past small shops selling aluminium pots and pans, postcards and carvings, or bright cottons of red and blue *lavalava* shirts printed with hibiscus flowers, exchanging the occasional *'talofa!'* with smiling passers-by. A cool coastal wind blew away most of the humidity, picking up a little white dust from the roadside as I walked past the Prime Minister's office and the immigration office at 'Ifi 'Ifi street. Aggie Grey's famous hotel lured me inside for a thirst-quenching beer among the green palms and tastefully traditional furnishings and traditionally tasteless tourists. Then I turned left into Falealili Street to begin the longish climb to Robert Louis Stevenson's resting place on Mount Vaea four miles inland.

It comes as something of a surprise to discover that the author of *Treasure Island* lies on the edge of Samoa's forests on

Mt Vaea overlooking Apia. *Tusitala*, the teller of tales, as Stevenson was known, bought the land for his home at Vailima in the same year as the terrible hurricane. Five years later he died of a stroke on the veranda of his house, which survives today as Government House. I trudged up the path cut by two hundred Samoans as they carried Stevenson up the hill to the site where he had wanted to be buried.

Sadly, I noticed that a few taro patches had been cut from the lower slopes of the Stevenson Reserve, eating into the forest. Higher up a small grey bird accompanied me, hopping busily around the branches, occasionally singing a pretty lilting song, then swinging its tail, spread like a small grey fan, to the left and right so vigorously that I thought it might twist itself off its perch. It was a Samoan fantail, unique to Upolu and Savai'i. I searched for its distinctive nest, woven from grasses and shaped like a cup with a long tapering tail, but found none. Higher up, the heat of the morning brought out rivulets of sweat on my brow as I scrambled over the red earth and twisted roots along the path. Then there was blue sky through the overhanging canopy and a last effort brought me out into a grassy clearing the size of a small terrace. In the centre was a simple tomb of white cement and on the side in bronze Stevenson's still-living words:

> Under the wide and starry sky,
> Dig the grave and let me lie.
> Glad did I live and gladly die
>   And I laid me down with a will
>
> This be the verse you grave for me:
> *Here he lies where he longs to be;*
> *Home is the sailor, home from the sea,*
>   *And the hunter home from the hill.*

The white paint on the tomb is scratched with graffiti now, but that can do nothing to diminish the place. The view is magnificent. Large forested hills sweep in from the right on the other side of Vaisigano valley. On the flat bottom majestic groves of palms dominate the encroaching aluminium roofs of Apia. Looking north I could see over the town to the sea and the white ring of surf outlining the reefs. In the flame trees beneath the neatly-cropped grassy platform, wattled honey-eaters searched for nectar in the clusters of flowers shaped like lobsters' claws.

Lying on my back in the warm grass above Vailima I wondered why so many writers and painters found so little happiness or inspiration in Paradise. The mystery of the islands has attracted many of them to Polynesia: Rupert Brooke, Herman Melville, Pierre Loti, Jack London, Somerset Maugham and of course Gauguin. It is almost as if the beauty of the surroundings saps the creative will. The natural world offers a parallel: some creatures, once vibrant and energetic colonizers, having landed on an island home suited to their needs, have evolved into simpler forms.

Darkness was falling about Stevenson's tomb before I thought of returning to Apia. New creatures emerged from the gloom of the hillside behind, gliding on monstrous blackened wings and turning into the breeze with such speed that their wing membranes vibrated loudly like the skins of kites. The flying foxes, on wings as wide as my arms, descended on to the croplands below to gorge themselves on papaya, guava, or the pollen of kapok flowers which open at night. This was a good spot from which to count their silhouettes against the sea as they emerged to feed. I watched them clumsily fluttering around the trees and methodically beating their way across the valley below, and wondered how they could have colonized Samoa so far out into the Pacific. Fruit bats cannot fly as well as birds, and rarely fly long distances from land.

There are stories of lizards and frogs falling on the islands from the sky after cyclones. Often there is an almost invisible rain of spiders, beetles and flies, pollen, seeds and spores that drift on the passing winds once the storm has passed. Most fall into the sea and die, but some reach new islands and evolve. It is much easier to reach islands if you have wings. After a passing cyclone, many birds and bats find themselves out of sight of land. Flying for days at a time is no difficulty for birds used to long migrations such as long-tailed cuckoos, which fly to the South Pacific Islands from New Zealand each year, or golden plovers which do so from North America, or even great seabirds such as the albatross, which wanders oceans at will. It has a wingspan of up to three and a half metres which enables it to fly everywhere in the Pacific but the Doldrums. It has no difficulty crossing water gaps to reach islands, though it may have a hard time finding a place secure enough to breed in; it therefore confines itself to the more remote atolls and islands.

Less elegant and more bulky is the giant petrel, which has a wingspan of about two metres. Its heavy bill and savage habits have earned this bird the name 'vulture of the seas'. It will occasionally land on the nest sites of other seabirds, seizing chicks from unwary parents as well as feeding on carrion. Unlike the smaller petrels and shearwaters, which feed on squid and crustaceans at the sea surface and breed on many of the islands of the tropical Pacific, the giant petrel is a rare visitor to these waters. But for birds used to forested valleys and sheltered bays, the open sea is a desert. These will fly until their last reserves of energy are used up – a day, three days, at most a week – then simply flutter into the sea and drown. Every mariner has encountered the friendly bird at sea which refuses to leave and seems quite unafraid; exhaustion is easily mistaken for tameness.

Most that manage to reach land will probably die there, without the correct food supply or without a mate. Eventually, however, a pair might arrive together, or a clutch of eggs might be fertile, and a flycatcher, kingfisher, pigeon or bat population can begin. Such a colonizing event might never happen again.

I journeyed across the mountains of Upolu to the O Le Pupu-Pu'e National Park on the southern coast. At 258 square kilometres, this is the only major forest park in the South Pacific. It runs from Upolu's highest volcanic peak, Mount Fito at 1,100 metres, down steep forest-covered slopes, growing over brown lava-field boulders to the coast. Here two of the world's largest bats, with wingspans approaching a metre and a half, are to be found. One species, named *Pteropus tonganus*, the Tongan fruit bat, after the island where it was first found, is in fact much more common in Samoa. These bats spend the day in colonial roosts high in trees, a habit which has earned them the local name of *Pea taulaga* or town bat. They are handsome creatures with fox-like faces, upright ears, deep-brown eyes, and manes of soft tan-coloured fur over their shoulders. The second type is known as *Pea vao*, the forest bat. This is supposed to have reddish fur over its shoulders but is otherwise similar to the Tongan bat. It enjoys the scientific name of *Pteropus samoensis*, the Samoan bat. This was the species I particularly wanted to see, as it is surrounded by mystery and confusion. Local mammalogists on the island had told me that it did not exist, and was merely a young Tongan bat. Some believed it to be rare and in danger of dying out; others thought it common. Most extraordinary of all, it lived alone in the forest and flew not at night but during the day. My first introduction to one was unexpected.

Outside his modern forestry department house, styled like a traditional Samoan *fale*, Tatua, head of the Samoan family with which I was staying, tossed some newly-skinned fruit bats on to the wire mesh over the fire. As I watched their leathery wings twist and shrink, he turned to me with a huge grin and said, 'You said you liked to see bats, so I shot you some.'

Tatua was to some extent a poacher turned gamekeeper – a hunter who was sent to Bulolo Forestry College in Papua New Guinea and returned to work in the Forestry Department in Samoa. Orange flames flickered over his hugely plump and almost naked body. From knees to hips, his legs were minutely decorated in fine indigo tattoos, precise geometric patterns depicting battles and great past events. To wound the body in this way takes weeks; the pain and blood endured is the mark of a true Samoan man.

Plucking a well-roasted bat from the flames, Tatua thrust two fingers under its ribs, scooped out the guts, and swept them into his mouth, closing his eyes with relish. Samoans regard these as a delicacy; being largely filled with pulpy fruit, they have a sweet flavour. He gestured to a carcass and said, 'Try some.' The meat tasted gamey, rather like roasted pigeon. It was tough but delicious.

Tatua loved to sing and dance and it was not long before he rose, cheered by the other Samoans around the fire, to bob and weave with arms outstretched like a bird of the forest, delicately changing the angle of his palms, then slapping his intricately decorated thighs and gazing up towards the stars, hooting and howling into the night. They insisted I dance too, and the pair of us revolved around the fire, crouching and stamping like a pair of alcohol-crazed demons. Inadvertently I stamped my foot on the burning embers, but filled with the energy of a Polynesian warrior, I never even noticed.

The following day we set off into the park past the beautiful Togitogiga Falls, a favourite picnic spot for Samoans, and into the forest scrabbling over the uneven lava boulders in search of bats. We found no roosts, but after several hours we arrived at a large cave in the pouring rain, where Tatua assured me there would be bats. It was a lava tube. Volcanoes do not always throw their lava into the sky. Sometimes it flows down tunnels beneath the mountainside. These 'lava tubes' are created when the molten rock hardens on the surface but continues to flow beneath. It is so hot that when the source ceases, the cavern empties, leaving a hollow tunnel sometimes many miles long, scoured with lines from ancient lava flows like rings around a bath.

I expected small cave-living insectivorous bats to be huddling in the roof, but instead found birds which behave like bats and are often mistaken for them; they were white-rumped swiftlets. These small birds can echolocate their way around caves, where they prefer to nest, in total darkness, using small clicks and twitters like bats. They make small cup-sized nests out of plant fibres glued together with their own saliva. A couple of small chicks snuggle inside each one on a lining of moss and lichens. In Asia the nests of their cousins are used to make bird's-nest soup, but those of the white-rumped swiftlet are considered too fibrous.

The bat roosts had eluded us, but that evening I was able to count more than 4,000 fruit bats flying across the road from the southern portion of the park to their feeding grounds in the north in the forty minutes just before dusk fell. Each evening this phenomenon repeats itself, as perhaps it has done for thousands of years. Some of the bats appeared to be flying south; they flew differently, were larger and had differently shaped wings: the elusive *Pea vao*? Suddenly shots began to split the night. The bats weaved and turned and some fell. Further away, the sound of gunshots, more frequent now, rang out along the road as far as I could hear. As the bats fell to the road, limbs broken and bleeding, the hunters grabbed their wings and swung them in an arc over their shoulders, smashing their heads on the road. The carnage continued for perhaps an hour until it was too dark, then the guns fell silent. The war on these harmless bats is not waged to provide food for the Samoan hunter's family, nor is it designed to prevent the bats from gorging themselves on the Samoan farmer's crops. It is to satisfy the traditional greed of gourmets thousands of miles away on the island of Guam.

Guam is the southernmost island in the chain known as the Mariana Islands in the Western Pacific region of Micronesia. The indigenous Chamorro islanders have been subjected to a succession of invaders – Spanish, German, Japanese, and finally American. The Americans currently own the island, and once used it as a base from which to bomb the paddy fields and jungles of Vietnam. The Chamorro, now mostly Americanized, have retained one aspect of their traditional culture: a love of bat soup. On feast days, fruit bats are simmered, fur, wings and all, in a broth of coconut milk and then eaten with relish. To Chamorro Indians, bats are a treat, and if you are poor you can even buy them on US welfare stamps.

The Second World War wiped out most of the forests on Guam, which is now covered largely with military bases, asphalt and huge creeping automobiles, so the supply of local bats was quickly used up. Hunters scoured the islands to the north so well that the Mariana Islands fruit bat, *Pteropus marianus*, is now on the endangered species list. Hungry for new killing grounds, the trade has crept across Micronesia, reducing the bat populations on numerous islands. As I stood in the dark beside the O Le Pupu-Pu'e National Park I knew that it had reached as far into the Pacific as Samoa, and that the bats there could now be doomed.

Thousands of bats are now illegally sent frozen from Samoa to gourmets in Guam in polystyrene boxes packed with ice. I asked Ray Tulafono, a senior member of the Forestry Department in Apia, about it and he told me that a law already exists to prevent the trade from Samoa, but that the Government has neither the resources, nor it seems the will, to prevent it. Middlemen provide free cartridges and a day's pay to the hunters who shoot the bats. They are given export licences and certificates of authenticity without which the bats cannot enter Guam. American Samoa also sends bats to Guam. Though there are supposed to be restrictions, no precise record is kept of the number exported. Hunters merely report the numbers shot, without providing proof. In Guam the bats will fetch anything up to $us35 each. At present there seem to be enough Tongan bats to withstand a limited trade; they thrive in the secondary forest and croplands which are gradually replacing Samoa's original jungle. How long will it be before these flying foxes are added to the list of the world's endangered species? And what of the mysterious Samoan bat, which depends on untouched forests?

This giant bat perhaps already merits that status, and to find it I had to travel to the last remaining area of extensive lowland forest left on the neighbouring volcanically active island of Savai'i. Surprisingly the ferry left on time from Mulifanua on Upolu's western tip. A couple of hours later we were docked at Salelologa and I took the long bus-ride along the southern coast, past the spectacular blowholes at Cape Asuisui where the surf bursts like geysers through fissures in the lava cliffs, sending rainbows across the sky. We turned inland from the palms and tangled overgrown taro gardens towards Asau Bay on the north side of the island. Several miles to the east of Asau lie huge raw lava fields, similar to those of Hawaii, remnants of the last great eruptions from Mount Elietoga and Mount Silisili in 1905 and 1911. The Faleolupo peninsula is at the north-western end of Savai'i. The forests there are some of the best I saw in the South Pacific Islands. The trees reached the height of twenty-storey buildings, and a number of the species are unique to this area. They have only recently been identified – by Paul Cox, an American researcher who has been studying the area for a number of years. He believes it to be the last potential stronghold of the Samoan flying fox.

From a hillside overlooking the undulating canopy of trees I was able to see these bats well for the first time. They fly quite differently from the other large species on the island, with fewer beats, and their shape is a less pronounced W. I could occasionally follow them with my binoculars as they spread giant wings over the forest in the sunlight, soaring on rising air thermals like birds of prey. Most bats do not soar; these were the most impressive I had ever seen. Elsewhere I found evidence which explained Paul Cox's concern. Plans to log this forest are well underway, supervised by representatives from

The Impact of Man

My journeys by canoe in the footsteps of the Polynesians brought me closer to understanding the supreme navigational skills of the early migrants. In the space of a few thousand years they spread themselves throughout the Pacific and became the most widespread people on earth.

1 **Langa Langa lagoon at sunset**: Today water still provides the highway on which most people travel between the islands.

2 **Tatooed warrior**: What drove the Polynesians to leave their homeland and search for new islands? Over-population, the struggle for power and a rise in the level of the sea may all have played a role in the distant past. These great events are still recorded today in body tattoos, the finest of which are to be found in Samoa and the Marquesas.

3 **Statues from a forgotten past**: Great stone monuments are all that survive of Easter Island's once magnificent civilization. The question of who built them and why remains one of the world's most controversial puzzles, although the answer is gradually becoming clearer.

4 **Beku idol**: The evidence of ancient cultures and civilizations lies scattered throughout the Pacific. This old wooden statue guards the skull houses at Pa Na Ghudu on Simbo Island.

5 **Skull shrine**: These ancient skulls date from a time when head-hunters roamed coastlines in search of easy trophies, and cannibalism in some island groups was a refined gastronomic art. On a sacred island in Roviana lagoon, in the Solomon Islands, a carved wooden door protects the spirits of the elders, whose skulls, following natural death, were placed there for safekeeping.

6 **Navala Village, Fiji**: Here village tradition survives but the landscape now is very different from that witnessed by the first migrants. Only a few trees on the skyline have withstood the annual burning of the hillsides. In the valley bottom, streams carry eroded soils to the sea.

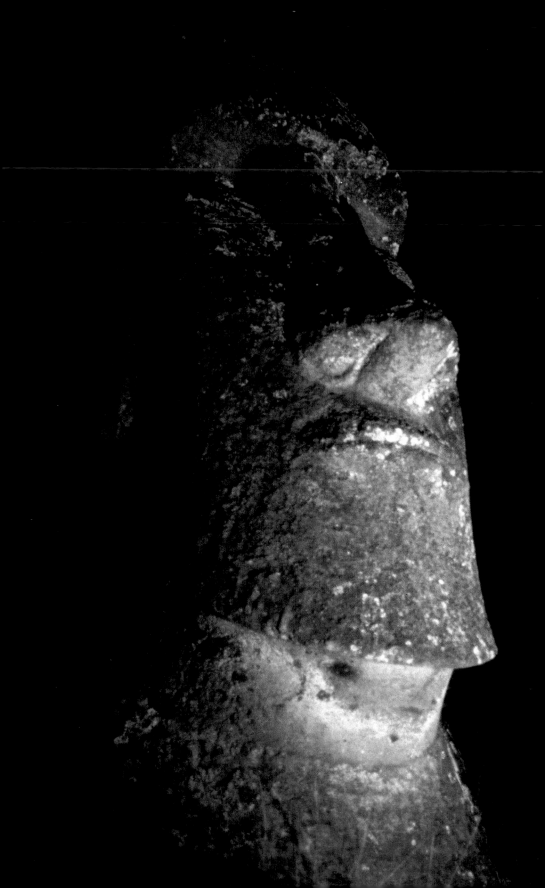

4

5

the New Zealand Forestry Service. I met none in Samoa who knew anything about the ecology of tropical forests or the role of a pigeon or a bat. Though meaning well, they can only see the forest in terms of timber management, mostly learnt in the temperate forests of New Zealand to which they would secretly like to return. Survey lines had been cut; chainsaw gangs were already at work. Moving along their tracks deeper into the forest was profoundly disturbing. Houses and banana plantations soon gave way to a wall of seemingly wild forest, but behind this a battleground was revealed. Trees sprawled drunkenly in all directions, stumps grey and twisted point-ing to the sky. All the larger trees had been felled. Those smaller specimens left standing appeared throttled with climbers and epiphytes which were racing up their trunks, feeding on the abundant light now available to them. On the ground there was new growth, but not of the original forest. Elephant taro, young banana palms, and coconut grew quickly on soil scheduled for new trees. Crushed branches littered the ground.

Landowners here, unlike those in many other areas, often do not want their forests cut. In March 1986 Salelologa's farmers burnt two million dollars' worth of logging equipment belonging to the South Korean South Pacific Development Corporation. They also dug up the graves of relatives of the Chief, who had capriciously signed away their trees against their wishes while living in Hawaii. In Faleolupo they were forced into an agree-ment to log when the Government threatened to close their old school for lack of maintenance. They needed $US25,000 for a new one, and the villagers could not raise it until the logging began. Only then would the bank give them a loan, guaranteed by the royalties they received (the bank would benefit addition-ally, of course, from deposits from the loggers). Not surprisingly the royalty offered by the Government precisely matched the funds required. The villagers relented for their children's sake and now the forests which they would have inherited are being destroyed, along with the unique creatures and trees they contain. Such tragedies are significant on a world scale. Savai'i's forests exist nowhere else. To remove the mix of species unique to an island is to remove them from the world altogether; there is no chance of importing them again from elsewhere. Samoa's remarkable bats, tooth-billed pigeon, friendly ground dove and forests teeter in the face of oblivion.

I visited one place in the Pacific where bats are safe by royal decree: the Kingdom of Tonga. There King Taufa'ahau Tupou IV (who, it is rumoured, has recently reduced his weight from 36 stone to a mere 25) rules with an almost feudal charm through Polynesia's oldest and only remaining monarchy. His dominions stretch across 1,000 kilometres of the Pacific from Niuafo'ou in the north to the ancient island of 'Eua in the south. Essentially the kingdom consists of three large groups of atolls, Vava'u to the north, then Ha'apai, then Tongatapu in the south. Only 37 of the 170 islands are inhabited. The first Polynesians arrived here more than 3,000 years ago, and developed into a powerful seafaring nation with an empire covering islands as far west as Rotuma, part of Fiji, and as far east as Samoa and Niue. The first European to reach Tongatapu was the Dutchman Abel Tasman in 1643. Cook followed 130 years later, presenting the Tu'i Tonga with a giant tortoise collected during *Endeavour*'s visit to the Galapagos Islands. It survived in the Royal Gardens at Nuku'alofa, the island's capital, until 1966. Overwhelmed by the hospitality he received, Cook named the islands the Friendly Islands.

Tongatapu is a small flat island shaped like a seahorse and barely thirty kilometres long. Two-thirds of the Tongan nation live here. Though there is little of the original forest left, the island supports a number of thriving Tongan flying-fox colonies and none is more spectacular than the one occupying the *Casuarina* trees down the main street of Kolovai village on the western tip of the island. The bats here are not shot at, so they are very tame and will not fly away if approached; indeed the whole hustle and bustle of village life continues beneath them while they watch with interest, twisting their chocolate-brown faces to and fro to watch buses and bicyclists, children and large Tongan ladies. It is a delight to see such harmony here. Despite the damage to the fruit crops that the bats undoubtedly cause, most people I spoke to welcomed them. They gave the village a certain amount of fame and brought in many valuable tourists. Watching the furry creatures from the roof of the local church, I was captivated by their flying skill as they came in to land on the topmost branches, often in strong winds. They would either fly over a branch, catching it with the claws on their hind feet, lurching to a stop like a fighter on a carrier's deck, or do a 180° turn as they approached and grasp the branch behind as expertly as a circus performer. Large mature males occupied the trees in the centre of the village, above a cemetery filled with sand graves. They spent most of the day attempting to mate with the females; a precarious occupation, the least danger of which is falling off your perch. A female bat who does not wish to be molested is a fearsome sight; she delivers a constant rain of blows and screeching rebuffs which one would have thought were sufficient to deter the most ardent suitor. Only when his loved one's wings are firmly gripped by his own can the male grasp her furry mane with his teeth and achieve success.

Less mature males and females were strung out along the high street to the east. The males here were no less determined but lacked the finesse of their elders, and seemed more easily put off. Those furthest away had given up all hope, and hung apparently asleep, wrapped behind dark leathery wings. In all I counted 6,500 bats in the colony, an increase of 1,500 over a census carried out four years before. Mafi, an old man of the village and the keeper of its ancient traditions, explained how these bats reached Kolovai. His story added another piece to the colonization jigsaw.

It seemed that many years ago, the King of Samoa at Apia invited the King of Tonga to participate in some games. The King sent his finest navigator, who excelled himself and was spotted by the Samoan King's daughter. She gave him a pair of bats and he returned with them to Tongatapu and kept them in his house. It so happened that the Chief of Kolovai was critically ill, and hearing that the bats had magical powers, he called for them and put them in a tree outside his house. Miraculously, he was cured, and the Chief asked that the bats should remain there, which they have done to this day. This story, which in essence is probably true, reveals that the Tongans were able to carry creatures long distances in their huge double-hulled canoes. In the Cook Islands I heard stories of how Tongans had brought bats there; elsewhere they had traded parrots and lorikeets across the sea for their scarlet feathers.

Man and nature were once again entwined in the flow of creatures across the Pacific archipelagos.

There is a gentle pace of life in Kolovai, an understanding perhaps of the value of nature to modern man. Tongans, like most Polynesians, bring a spirit of enjoyment to life; there is always time to play cricket amongst the palms, to sing a song, to admire the rainbow colours of fish brought ashore as the sun sets. Nowhere is this demonstrated better than in their delight in feasts. These are a celebration of abundance and good nature. Their preparation binds villages to their King; they are to be admired as well as consumed. Such a feast, once attended, is never forgotten.

Great processions of lorries trundled towards the eastern tip of the island. Each contained four or five roasted pigs, nose to tail on a trestle shrouded behind white lace, supported on arches of smooth branches. The pigs' feet were decorated for the occasion with small pyramids of taro, and on to this were placed baked crabs, boiled crayfish blushing along their backs, fish, taro leaves, tinned meats, bananas, coconuts and sweet potatoes. At the feasting ground, a square of leaf shelters had been constructed for the thousands of guests. Schoolchildren enacted vigorous *kailao* war dances in skirts of long leaves mottled in red and green, rings of nuts rattling around their ankles, white plumes dancing on their heads as they stamped to the beat of drums and thrust a hundred spears up into the perfect blue sky. With shrieks of delight vast Tongan women rushed forward and thrust *pa'anga* notes into the dancers' shorts; young men stuck them to the girls' oiled bodies.

The trestles of roasted meat and trimmings were set out; there were 2,500 pigs awaiting the King's arrival. He emerged finally from a large blue sedan, in a knee-length black leather overcoat and a pair of skiing goggles. He walked slightly unsteadily on a walking stick to his pavilion, and sat cross-legged on the ground behind a table burdened with food fit for twenty kings. He had just opened the college's new chapel and was feeling a trifle hungry.

Chapter Seven

EVOLVING ALONE

Hawaii

> The loveliest fleet of islands that lies anchored
> in any ocean.
> > Mark Twain, *More Tramps Abroad*, 1907

ne legend in ancient Hawaii tells of the day that Maui set out to capture the sun. Before sunrise the Polynesian demigod crept to the summit of the giant volcanic crater of Haleakala and lay in wait with fifteen coconut-fibre ropes. He was used to great feats, and had a taste for mischief. Earlier he had raised the sky from the earth so that people would no longer have to crawl on their bellies and leaves would no longer have to be flat. The sun was then able to come up from behind the world and run across the sky on his many fiery legs.

Maui's ungrateful mother Hina had complained that the days were too short for her to beat tapa cloth and gather taro because the sun ran too fast, so as the first glistening legs of the sun reached over the crater's rim, Maui lassoed each one and secured them to a williwilli tree. He had trapped the sun. He attacked it with a magical bone until it pleaded to be released. Only when the impatient sun promised to move more slowly and give more time for the beating of tapa cloth and the fishing of reefs did Maui hint that he might release it. Even so he left some ropes attached for good measure, and each day as the sun sets these can still be seen trailing across the sky. Haleakala is known to Hawaiians as the House of the Sun.

In the darkness before dawn, on the sharp edge of Haleakala, it can be very cold. At 1,800 metres on Maui in the Hawaiian Islands in the middle of the Pacific, the wind cuts into exposed

skin and creeps in through the openings in clothes to touch
flesh like the cold blade of a knife, an odd sensation on a
tropical island. Beneath my feet the stones were shattered and
chipped like slates, clattering as I moved position to keep
warm in the face of the buffeting wind. In the distance far
ahead, black silhouettes gradually began to form: jagged
peaks with rugged shoulders, the crater's rim set into a back-
ground of slowly fading stars.

The sun edged over the crater's rim. Haleakala was suddenly
filled with golden light, its hues and outlines changing as the
minerals in its shattered rocks flashed and sparkled like crys-
tals: a crater filled with jewels. Sweeping shingle slopes
changed through shades of red, brown and gold between black-

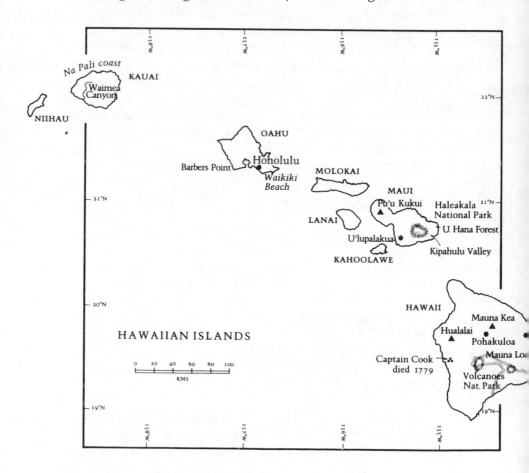

ened outpourings of once-molten rock. For the first time I gazed into Haleakala and drew breath at its stark and unrelenting beauty.

Water, wind and fire have joined forces with the sun to fashion one of Hawaii's greatest natural wonders. The giant crater of Haleakala dominates the centre of the island of Maui. From where I stood it was 900 metres to the bottom and 12 kilometres to the rim opposite. In total it encircles 45 square kilometres of some of the most spectacular scenery in the world. As the sun rose massive cinder cones emerged from the gloom, their smooth slopes swathed in morning mists that hung in the still air, like islands in a celestial sea. Dark vents which once spouted tall curtains of fire were now silent and lifeless; the last eruption was in 1790, when boiling lava poured down the slopes outside the crater and into La Pérouse Bay. Haleakala is not dead, but dormant. The whole area below was a moonscape of browns splashed with red and gold. Barren, desert-like, it seemed there would be little that could survive in this wasteland, yet here within the crater, and nowhere else, some of nature's most extraordinary works have evolved. I loaded my small backpack on to my shoulders and started to descend into the crater down the Sliding Sands Trail.

Sabre-toothed tigers and woolly mammoths still roamed the earth at the time Maui began to push up from the ocean floor. The group of islands to which it belongs is a conveyor belt of gradually ageing volcanoes, the youngest in the east, the oldest in the west. As with Rarotonga in the Cook Islands, their origins lie beneath the sea at a 'hot spot' in the earth's mantle from which bright lava spills into the deep. The immense pressure and cold at that depth prevent gases in the lava from escaping and it spreads flat across the ocean floor like molten red toffee. Maui's first great building period, or Hanomanu as it is called by geologists, began a million years ago. Thousands of years later, as the sub-sea mountain peak neared the surface, super-heated lava caused giant explosions of steam which sent great ash clouds high into the sky.

The young island continued to grow as lava poured from its summit and won the battle with the sea. Haleakala now rises to a height of 3,055 metres above the surface but its slopes descend a further 5,791 metres to the ocean floor. As it grew it fused with a second, older volcano, Pu'u Kukui, which is now known as West Maui. The whole island of Maui was at one time much larger than it is today but the other volcanic peaks of 'Maui Nui', Big Maui, have been cut off from the present island by a rise in the sea level since the last ice age. It now covers an area of 1,800 square kilometres.

The path was steep and slippery with sand and cinders. My descent was a lesson in geological beauty. The walls of the crater are composed of dark erosion-resistant basalt, much of it forced through the more ancient lava flows as intrusions of black stone. The finest-grained basalts were highly prized by early Hawaiians for making stone axes; these could be used to fashion canoes. This Kula period of volcanic activity resulted in the huge cinder cones, now much eroded, as well as rugged dykes and plugs surrounding the crater rim. I looked up to the point at which I had begun. At the end of the Kula period, Haleakala had been 900 metres higher.

A high mountain in the path of moisture-laden Pacific trade winds attracts rain. That which fell into Haleakala carved valleys and spread sands, etching its crags and dry-stream channels. Outside the crater great sweeps of alluvium washed from the volcanic dust provided soil for plants to take hold. Waterfalls ate backwards up the slope, carving valleys which eventually broke into the crater at a number of places around its rim. My shoes and jeans were covered in volcanic dust. The sun was higher now, and hot. It seemed that there was still a long way to weave through the huge lava boulders towards the crater floor. On either side of the path there were small plants growing from the brown rubble of stones. I was surprised to see a dandelion; perhaps its parachute seed had been brought in from America attached to someone's shoe, for it is not a native of Hawaii. It was strangely comforting to see something so familiar in such a desolate place.

Other plants have been here much longer. The harsh sun and ultraviolet light at this altitude, coupled with dry conditions due to water quickly percolating through the stones, make for difficult living conditions. The *kupaoa* is capable of withstanding the sun's intense radiation and the lack of moisture because its leaves are tough and spiky like those of a monkey puzzle tree. But there was another strange plant I was especially keen to see, the largest specimens of which are found only in the crater of Haleakala.

Reaching the crater bottom I rounded some boulders and there they were, appearing like a collection of giant pincushions among the rocks. Huge dried flower spikes rose well above my head from the balls of silvery leaves at their bases; they are known as silverswords. Peering closer I noticed that the leaves were covered in fine white hairs to protect them from the sun and retain water from the mists which fill the crater at night. The growing bud is concealed deep within its protective pincushion as frosts are not unknown on Haleakala.

The silverswords' ancestors appear to be related to *Dubautia*, the Naenae plants, which are similar to the tar weeds of California. They may have evolved from a single early introduction to the islands. The Naenae evolved into shrubs and even trees and vines. Those in the mountains became the silverswords, with different varieties adapted to cold, dry or moist conditions. The Kau silversword has short woody stems as it has no need to fear frosts on the ridges of Mauna Loa on Hawaii, where it grows. The species which grows on Pu'u Kukui in West Maui is subjected to intense rain and has evolved flowers that are upside-down so as not to fill with water. The *iliau*, another relative, grows a long stem which holds its leaves above those of surrounding shrubs to enable it to reach the sunlight it needs.

The heat in the crater bottom was now almost unbearable. I stopped to eat in the shade of a massive boulder. In the distance some tourists rode across the crater, shimmering like a camel train.

The huge cinder cones were now not far away, much bigger than they had looked from the rim. These were remnants of Haleakala's last eruptive era, the Hana period, which had ended almost a thousand years before. I looked at the cones and imagined the fire fountains, the roaring and the smell of sulphur that would have filled the crater then. As the hot lava fell to earth it cooled, landing as a rain of solid boulders which I could see scattered around. Those that had not cooled before landing remained soft and became like stone cowpats. The majority of ash had fallen on the western side of the cones; the wind must have blown from the east all those years ago, so American seeds could perhaps have been blown here.

The area between the cones was tempered with hues of red and brown and is known as 'Pele's paint-pot' after the Hawaiian goddess of fire who lives there and who Hawaiians believe is responsible for creating the islands with her fiery breath. Near to the centre of Haleakala there is a huge pit, Keanawilinau, where Pele's sister Kamakokahai, goddess of the ocean surface, tried to quench Pele's fire by forcing her way into the crater. In ancient times the umbilical cords of newborn babies were thrown into the pit in the belief that this would prevent them from becoming thieves.

It was time for me to leave Haleakala. The climb back out of the crater was exhausting in the searing heat. The rocks were too hot to touch, and the whole area shimmered like a desert. It was hard to believe that outside the crater in the face of the wind there would be water in abundance. A winding road of asphalt dropped down from the peak through the shrub and tree zones that clothed the outer flanks of the volcano. At 3,000 metres the slopes were barren and lifeless. Lower down shrubs began to appear sprinkled with the orange berries of *pilo* bushes, food for mountain birds. Suddenly I entered a zone of thick cloud, the inversion layer, where cold air above meets the warm air below and turns it to mist, a welcome wet zone on such Pacific mountains which dampens the leaves of thirsty plants. The cloud forests which face the eastern winds harbour the great variety of plants on Maui. The dry forests on the opposite side are completely different, with their own characteristic species. Such variety on an island so isolated in the sea is remarkable.

Evolution moves in strange ways on islands far from land. Nowhere has it been stranger or more enthralling than in the Hawaiian Islands. This is the most isolated island group in the world, situated in the northern Pacific, 4,000 kilometres from the nearest major landmass and 1,600 kilometres from the closest island groups. Maui is one of eight high volcanic islands in the young eastern end of the chain stretching across 585 kilometres of ocean from Hawaii to Kauai. The Hawaiian chain

does not end there but continues west, as a series of progressively older pinnacles and atolls, for a further 4,830 kilometres
to Midway, where it bends sharply north, continuing beneath
the ocean as the Emperor Seamounts which finally vanish into
the earth's crust at the Kuril-Kamchatka Trench. These dots
across the Pacific map are the quintessence of island evolution,
beginning with the volcanic birth of young islands in the east
which are then carried on the conveyor belt of the Pacific plate
to die in the west like a string of gradually sinking Noah's arks.
It is an evolution which began fifty million years ago and is
continuing. A new island, Loihi, is slowly rising beneath the
sea from the hot spot which spawned them all at the eastern-
most tip of the chain, and will one day take its place in the
Pacific's greatest island arc.

To see the Hawaiian Islands from the air is a magnificent
experience. Beginning in the east the first island to appear is
boisterous and smouldering Hawaii. It is the youngest of the
islands, at a little under 700,000 years old, but it is also the
largest and therefore known as the 'Big Island'. As you pass over
the Puna coast, black sands tell of lava once spilled into the sea.
Steam rises as the ocean boils where a new lava flow trickles
into the sea from a small active cone rising like a suppurating
boil on the massive smooth flanks of Mauna Loa volcano. The
brightness of the orange molten lava stands out against the
black solidified rocks through which it runs, occasionally
disappearing from sight as it dives beneath the surface into a
lava tube.

To the south-east the lava wells up into a molten reservoir, a
lake of burning orange which bubbles and spits at 1,200°C. Two
small helicopters hover like flies around its edge, giving tourists a closer look. At one edge the lava surges, splits its thin
black skin and suddenly flows downwards, disappearing again
to emerge nearer the coast where it has cut off the road to
Pu'uloa. Burnt-out houses and cars lie scattered in its path
alongside the grey remains of trees. The present eruptions have
continued for months.

Further north, the massive crater of Kilauea looms ahead, a
shallow circular scoop out of the mountainside more than
three kilometres across and one of the most active craters in
the world. Inside is the smaller Halemaumau 'fire-pit', home of

the Hawaiian goddess Pele, famous for her great curtains of orange molten rock. As you leave the summit of Mauna Loa to the right, the ground falls away across the saddle which joins it to its less active sister, Mauna Kea. The summits of both the two highest mountains are often snow-covered, but Mauna Kea is the higher and indeed claims a title as the biggest mountain in the world, reaching 10,204 metres from the ocean floor. It is still active and growing. Grasslands grazed by cattle now cover the once-molten lava flows. To the west the white sands of the Kona coast testify to the corals which now fringe the coast there.

Leaving the north-western tip of Hawaii behind, it is another fifty kilometres to the twin peaks of Maui and the two smaller islands of Kahoolawe and Lanai, which were once part of it; the valleys which joined them have now sunk beneath the sea. Near the cloud-fringed summit of East Maui, a moonscape world opens up beneath you of jagged brown cliffs and treacherous screes, the giant caldera of Haleakala. Lava flows a mere two centuries old stand out on the mountain's slopes, still poorly covered with trees.

You sweep across the pineapple-covered lowlands, over the old whaling town of Lahaina, where the first kings to rule all Hawaii once lived, and out into the ocean again. Molokai is easily visible to the right, showing off the fluted green cliffs of its Pali coast, the highest sea cliffs in the world, falling a thousand metres to the ocean. Soon Oahu, capital and commercial centre of the Hawaiian islands, looms ahead. Layers of thick white and grey clouds crowd the twin mountain ranges which flank its central valley; mountains which are more than three million years old. Some of the finest surfing in the world is to be found along its windward coast. And to the south lies the setting for one of the world's greatest tragedies: Pearl Harbor slides beneath the aircraft's wings. Behind, the tower blocks which now crowd Waikiki blend into Honolulu, scrambling like a white fungus into the highlands. It is another sixty kilometres to the ancient island of Kauai and its small privately-owned neighbour Niihau.

The Hawaiian Islands harbour about 1,000 species of flowering plants, 2,000 of lower plants, 8,000 insects, 1,000 land snails, 1,500 marine molluscs, over 100 birds, 680 fish, 3 sea turtles, a bat and a seal. Almost a quarter of Hawaii's flowering plants seem to have their origins in the Americas; the rest probably originated in the far north and far south of the Pacific, with the majority coming from South-east Asia. The variety of animals is not great considering that the land area they inhabit exceeds 16,000 square kilometres, but almost 95 per cent of them are not found anywhere else. The islands provide an extraordinary range of habitats which evolution has explored, exploiting the genetic information that drifted or was carried or blown on to their shores. Colonizers could stake their claims on molten rocks hardened deep beneath the sea or the summits of snow-covered mountains at 4,200 metres. They could find refuge in dunes drifting along beaches, arid lowland forests, high-elevation swamps, the canopies of rain-soaked tropical trees abundant with clinging epiphytes, coral pools or subterranean caves. The islands are a melting pot where evolution has run riot, where single colonizing weeds have evolved into beautiful trees, where a few fugitive birds have been transformed into many, whose grace and brilliance of colouring are unsurpassed.

Ever since human beings first set foot on Hawaii's majestic coasts, these islands have entranced them. The handsome women explorers found inhabiting Hawaiian shores have tormented the souls of Europeans ever since Cook arrived 211 years ago.

Karakakooa Bay, where Cook unwittingly landed on 17 January 1779, was known to the Hawaiians as the birthplace of Orongo. The Hawaiians believed their god had returned and crowned Cook eight days later in an elaborate ceremony.

Cook's ships left on 4 February, amid heartbreaking farewells. Three days later, following storm damage, they returned for repairs. The confused natives pestered the ships, for their beliefs had been undermined. Uncharacteristically Cook lost his temper, went ashore with a party of marines to demand the return of a stolen cutter, and took Chief Terreeoboo hostage. In the mêlée some marines shot a high chief escaping in a canoe from boats in the bay. Incensed, the Hawaiians attacked. Aboard the *Resolution* a horrified Lieutenant King recorded the end:

The islanders, contrary to everyone's expectations, stood the fire with great firmness; and before the marines had time to reload, they broke in upon them with dreadful shouts and yells.

Our unfortunate Commander, the last time he was seen distinctly, was standing at the water's edge, and calling out to the boats to cease firing and to pull in . . . whilst he faced the natives, none of them offered him any violence, but having turned about, to give his orders to the boats, he was stabbed in the back and fell with his face into the water.

Over several days priests, hopeful of restoring peace, returned his dismembered body. Some bones were still being revered in rituals thirty years later when the first missionaries arrived. Today a lonely white pinnacle marks the place where Cook fell on the Kona coast.

The Polynesians preceded Captain Cook by some 1,500 years, journeying in their massive double-hulled canoes from the Marquesas, and later from the Society Islands. They found the islands already occupied by an extraordinarily rich variety of plants, from coastal morning glories, to handsome ohi'a trees festooned with ferns and orchids. Numerous birds they had never seen before filled the valleys with sweet song, and pale-green butterflies and dragonflies paused at streams, while by night hundreds of moths drank nectar from white flowers. No mosquitoes plagued their skins, but numerous unusual flies displayed their patterned wings on ferns, and in the bushes they found such a variety of delicately painted snails that it was impossible to count all the different forms. In the mountains there were large flightless geese, and small rails crept about the undergrowth, also apparently unable to fly.

There were some things missing which they had expected to see. The plump pigeons they so enjoyed eating were absent; there were no large fish in the streams; no coconut palms fringed the coast; no tasty lizards scuttled over the leaves in the forest; they heard no 'tchuk-tchuk' of geckos in their houses at night. The colourful lorikeets, whose feathers they so prized for decorating fine mats and clothes, also seemed to be absent. There were a mere two or three plants they could use for food; most that they knew from their homelands did not seem to grow here at all.

The first animals and plants to arrive on the Hawaiian

Islands will never be known as they and their fossils have been carried far beneath the sea, but today on the surviving high islands of the eastern part of the chain, despite their extreme isolation from the continents, the descendants of an extraordinarily diverse collection of animals and plants are to be found. The flora and fauna evolved here in an environment with few predators and little competition from similarly equipped creatures which they had left behind in their countries of origin. The first plants to arrive on these deserted beaches found themselves castaways on a shore filled with opportunity. Mountains and valleys which perhaps only they were in a position to colonize stretched before them. The animals which first trod the cooled lava boulders were likewise released from factors which may have held them in check elsewhere, and their genetic potential ran amok. Such was their isolation that they might be the only bat or bird or fly ever to reach that shore.

Had Darwin called at Hawaii after his historic visit to Galapagos aboard the HMS *Beagle* in 1835, he would hardly have been able to contain his excitement, for he would have seen examples of evolution which would have made those on Galapagos pale by comparison. Tiny flies have grown into giants, daisies have turned into shrubs, crickets have gone blind, garden lobelias have evolved into trees, caterpillars have become carnivorous, geese have ceased to fly, and some small birds have even become vampires. There are numerous species of honey-creepers, many brilliantly-coloured, but not all of them eat honey. Some prefer grubs prised from bark and have beaks for the job like woodpeckers, while others have evolved to eat seeds and have tough beaks like finches. Where did all these strange creatures originally come from? How did they manage to reach this isolated archipelago, and when did they arrive? How did the genetic information contained in their chromosomes change to meet the new conditions and selective pressures they encountered? In short, how do new species arise and how long does evolution really take?

Nowhere in the world is there a better place to study this than in Hawaii; here it is possible to watch evolution simplified, as it was for Darwin in Galapagos. The islands he found were isolated, young and hostile. Time had perhaps not allowed many plants or animals to arrive, and few of those that did were able to cope with the harsh surroundings. Among the finches that had succeeded, Darwin could see that the adaptations each possessed to exploit a certain food, their different plumages or body shapes, were sufficient to distinguish one from another, but that nonetheless they were all closely related. That much was easy, but to interpret the discovery to arrive at an understanding of how such evolution occurs remains the greatest feat of any biologist. In his diary at the time the first inklings of the magnitude of his discovery were forming in Darwin's mind:

> Hence both in space and time, we seem to be brought somewhat near to that great fact – that mystery of mysteries – the first appearance of new beings on this earth.

In Hawaii many continue Darwin's work, but today they are equipped with a modern perspective of ecology, molecular tools and powerful computers to run their statistical programmes. One discovery involves a tiny fly known as *Drosophila*. This is commonly known as the fruit fly for its habit of descending in large and rather annoying numbers on pawpaw during picnics or on fruit crops. To biologists it is of particular interest as it has enormous chromosomes, the genes of which are relatively easy to map. Whenever I have watched these flies on my pawpaw they seem to be in a state of copulation, numerous small males riding about on the backs of females. New generations are produced rapidly, so the effects of any manipulation of their genes by an evolutionary biologist are quickly revealed in the offspring. These little red-eyed black flies have become models for studying evolutionary biology as a result.

There are about 3,000 species of *Drosophila* in the world. In Hawaii, the land area of which is smaller than the state of Massachusetts, 500 have been described and possibly as many again are waiting to be discovered. They occupy disparate niches, from morning glory flowers on the coast to the slimy exudate of trees as high as 2,000 metres on Mauna Kea. They signal to each other in semaphore on tree ferns in rainforest, or scramble over desert-like sands on dry cliffs. They fill every conceivable ecological niche a fly could occupy so that now there is a multitude of species, evolved from one or at the most two original colonizers.

A large proportion of the insects found in Hawaii appear to have had their ancestors in Indo-China. While birds may be able to fly long distances over oceans, these flies could not. Storms sweep numerous insects and other small creatures into the upper atmosphere where they come into contact with the jet stream, a column of rapidly moving wind high up in the earth's atmosphere. Aircraft use this to quicken their passage across the Pacific and so have Hawaii's insects. The jet stream travels at anything up to 150 kilometres an hour and at that speed the floating insects could reach Hawaii in a matter of days from America or Indo-China. Many are hardy and seem able to survive the cold temperatures of the upper atmosphere for that short time. One jet stream from Indo-China slows down over Hawaii and insects could simply float down on to the islands. This may explain why so many of the insects found here have their nearest relatives in Japan, Korea and Vietnam. Nets used on ships at sea or on aircraft have found that an aerial plankton floats on the wind over the Pacific filled with young spiders on parachutes of silk together with many small flies and other insects. The ancestors of Hawaii's pomace flies were probably blown there in this way.

They found an island deserted of flies but rich in habitats waiting to be exploited. There were few birds skilled at eating insects, if indeed there were any at all; nor were there a great many spiders or other insect predators. Evolution proceeded unconstrained. The flies, normally a few millimetres long, have become ten times as large. Varieties that would otherwise have been weeded out by natural selection have survived here and new species have proliferated. With so many species to choose from, it was important to recognize the right mate and maintain the integrity of the gene pool, maximizing benefits passed on to new generations; so elaborate courtship rituals began to develop which are unknown anywhere else in the world.

The male pomace fly uses the same courtship system as the bird of paradise in New Guinea and the cock-of-the-rock of North America. Cocks of these species display to each other in special arenas or 'leks', hoping to attract females and knowing that the finest display will win the day. Male pomace flies do exactly the same, though for them the lek is not an open piece of grassland, as for the cock-of-the-rock but a collection of carefully chosen branches where they will be most visible, as for the bird of paradise. A male pomace fly positions on a favourite branch, fern frond or leaf and defends his territory vigorously. The males of another species have evolved heads like mallets with which they beat each other off branches. Males of *D. nigribasis* are more sophisticated. They face each other standing on tiptoe, wings pointing to the sky, and rock menacingly from side to side, touching alternate wing tips in a delightful slow motion kung fu. Some species have an intricate lacework of black lines over their wings and are known as picture-wing flies. A male uses the wings to attract females, waving first the left and then the right. If he is successful, he moves closer and allows the female to kiss his tongue with her proboscis, then he arches his abdomen over his shoulders and showers her with aphrodisiac perfumes from special scent-dispersing brushes in his tail: behaviour reminiscent of butter-flies. Then the act is consummated.

Recently a discovery has been made which confirms pomace flies as some of the most extraordinary in the world. They talk to each other. Ken Kaneshiro, a biologist at the University of Hawaii at Honolulu, has done a great deal of pioneering work on the flies. I went to visit him in his lab on Oahu Island, where inside small clear plastic boxes he had colonies of different pomace flies. In Hawaii there are not many juicy fruits to eat so the flies have evolved to feed on sap oozing from branches or decaying flowers and leaves. One has turned carnivorous and eats spiders' eggs. But the numerous species are distinguished by where they lay their eggs rather than by the foods they eat. Most of them will eat lots of foods, but will only lay eggs in the base of one species of tree or on just one fungus. Some of them will breed only on mushrooms; others prefer bananas. One species lays its eggs in the slimy flux on the branches of *Myoporum* trees, while another lays on the ground where the drips land. The flies differ only in a single gene, but this has totally changed their behaviour. It seems that without many predators or much competition, sexual behaviour has deter-mined the course of evolution in these flies; and because it is

clear that a small genetic change can have a big effect, it could
be that evolution has occurred much faster than was thought.

Kaneshiro moved a small piece of equipment into one of the
boxes and switched on an amplifier in a bank of electronic
equipment. A series of 'pfut-pfut-pfut' noises came out of the
speakers, followed by a purring nose not unlike that of a cat.
One pair of flies were displaying near the small microphone.
Suddenly one appeared to vibrate its abdomen and a series of
'pfuts' came out of the speaker. I looked at Ken amazed; he
smiled. 'They purr with their wings too, when they're trying to
attract each other. Watch out for the males when they get mad!'
A third fly approached the courting pair and one turned to face
the aggressor, signalling with its wings. The speaker emitted a
sound which can only be described as a growl. The interloper
was repulsed and the happy couple resumed their courtship.
Kaneshiro believes that three undiscovered sound-producing
organs must create these sounds which are quite different from
those of any other insect and are emitted at frequencies as high
as 15,000 Hz. So far he has tested a mere nineteen of the
thousand or so species of *Drosophila* he believes to be on the
islands.

Evolution of insects in the Hawaiian Islands has turned up
many peculiarities. Hawaiian damselflies lay their eggs not in
streams but in water trapped in the leaf axils of plants growing
up in tree branches. Their nymphs do not content themselves
growing there but prowl over the surface of leaves in search of
prey and even creep along decaying leaves. The genus of *Mega-
lagrion* damselflies has radiated into as many different species
of dragonflies and damselflies as are found in the rest of the
world. The bug *Saldula* which normally lives in rivers has in
Hawaii taken to the trees. *Euphorbia* bushes produce a sticky
white latex to gum up insect jaws but a Hawaiian leaf-hopper
has evolved a way of digesting it by vastly extending its gut into
a horn sticking out of its head like a unicorn. None of its
relatives elsewhere in the world has developed this awesome
equipment. Some creatures have undertaken what are called
'adaptive shifts', and have specialized in eating a food quite
alien to their ancestors. One bug which normally eats seeds has
evolved to eat dead insects instead. Most extraordinary of all
are the caterpillars of Hawaiian *Eupithecia* moths. A few years

ago these were found lurking on leaves and twigs with their forelegs raised in the air waiting to ambush passing flies and spiders, even crickets and wasps. They have evolved into beasts with huge talons and will snatch anything they can reach, devouring it with jaws most caterpillars use to chomp mildly away at leaves. This appetite for meat is unknown in any other of the world's 100,000 kinds of caterpillar. Little has been done to explore Hawaii's spiders. One wonders what denizens await in addition to the blind no-eyed-big-eyed spider which has been discovered lurking in Hawaii's lava tubes.

Gigantism is a common by-product of evolution on islands, as in the case of the longhorn beetles of Fiji, some of which are as large as dinner plates. Often this is as a result of adapting to eat large food items. The dragonflies and some of the grass-hoppers in Hawaii are among the largest in the world.

Flightlessness is another feature of many island species in the Pacific. Quite why it should occur remains a mystery. Perhaps it is simply that once a castaway has reached an island, it would make no sense accidentally to leave it again. An insect or bird is prone to be swept on the wind out to sea, where it might drown. Adults may have evolved progressively to retain the features of their young, whose small wings, weak muscles and unhardened breast bones prevented them from flight. The absence of predators such as weasels, foxes or slithering snakes on Pacific islands, together with the fact that plenty of food is available on the ground, are also possible causes. Nowhere has this occurred more spectacularly than in New Zealand, where the range of flightless birds that have evolved, from moas to kiwis and kakapos, is unparalleled. Sadly, flightlessness can spell doom for some birds. Before man arrived, Hawaii was certainly home to a large flightless goose, and the world's only flightless ibis once probed the marshes and shallows of the island coasts.

Plants have followed a similar course. Many which had lightweight floating seeds designed for dispersal by the wind or on the sea have evolved on Pacific islands to produce larger ones. These can store more food for offspring, as well as keeping them more firmly anchored to the ground, close to the favour-able conditions found by their parents. Many plants such as the beggar tick have spikes on their seeds which catch the wings of birds. The seeds of the pepper weed *Lepidium* used to be sticky, but lost this property once the plants' ancestors established themselves on islands. There were no mammals, nor birds initially, to disperse seeds, so stickiness was no longer useful.

When I first arrived in Hawaii I was met by the welcoming smile and southern drawl of Paul Atkins, with whom I was going to stay. He is a cameraman with unsurpassed experience of making magnificent wildlife films in the Pacific.

After the discomfort of canoe rides in the Solomons and thatched roofs in Fiji, Honolulu really did seem like a concrete jungle. A million tourists a year pass through here. There is time enough only to pause at the Hawaiian pearl-oyster stall (they come from Japan), down a giant Coke, fall off a surfboard at Waikiki and take an 'unbelievable' sunset cruise aboard the huge motorized canoe before boarding the micro bus to the 'unforgettable' Hawaiian cultural show, complete with the sensuous rustle of plastic grass skirts and electric guitars. The Hawaiians were consumed by first colonialism and then consumerism; their blood spread through Caucasians, Japanese, Filipinos and Chinese; their native skills, like their natural world, were swept aside by free enterprise. They are a people forgotten in their own land, clinging to the edge of extinction.

To discover some of Hawaii's greatest secrets of evolution it is necessary to travel inland. There is little flat land on most high islands of the Pacific. In Hawaii this has kept people to the coasts and most modern development is concentrated there. Even a short distance inland the world is transformed into beautiful valleys filled with tree ferns and majestic *koa* and *ohi'a* trees (a kind of myrtle) sporting red blossoms. From these forests the sound of native birdsong floats on the air and among their sheltering branches all manner of strange plants and animals can be found. To get close to them involves a drive or hike into the mountains; this is easy on many of the islands though most of their deep interiors are still rarely visited. Fortunately, the Hawaiian Islands follow the United States' enviable system of National Parks, though even in these the animals are not always safe. On the island of Maui, near the eastern end of the Haleakala National Park, there is a valley called Kipahulu. The upper reaches of this contain some of the finest rainforest left in Hawaii. As such it is reserved for nature alone and few humans other than scientists may visit it. With the kind permission of Ron Nagata, Chief of Resources Management at the Park, I flew there by helicopter to see what they were doing.

We burst through a thin layer of flat cloud and headed around the northern side of the mountains, skimming the tops of ironwood trees. The centres of these *ohi'a* trees are incredibly hard; ancient Hawaiians favoured them for the making of house posts and simple furniture. They dominate the landscape of the uncultivated slopes of Hawaii's mountains, offering a multitude of spiky red blooms to nectar-foraging birds and bees which have specialized in using them, pollinating the trees in return. The trees produce especially lightweight seeds which blow on the wind and probably account for their widespread distribution in the Pacific. They are excellent pioneers and will colonize lava fields soon after they have cooled. They are remarkably adaptable and come in many forms, occupying the hills everywhere from the warm lowlands to frosty summits; hence their scientific name, *Metrosideros polymorpha*. Stranger plants were waiting in the bogs below.

Mists were beginning to shroud Haleakala's eastern craggy slopes as we dropped through a gap into an amphitheatre of rocks several hundred metres across. I leapt out and squelched through the swamp followed by Art Madieros, a scientist with the Parks Department. The helicopter rose through the mists and vanished. An eerie silence filled the place: there was no sound at all save the whispering wind. The view was consumed by billowing mists and it was cold. We stood in an ancient crater surrounded by dark-green shrubby vegetation. Rounded tufts of herbs and grasses filled its bottom like a vast collection of hairy marshmallows.

'Watch where you put your feet,' Art said. 'Almost everything growing here is either endemic to Hawaii or endangered.' I looked aghast at the footprints I had made across the bog and wondered what priceless examples of evolution I had already consigned to oblivion.

'How do you expect me to move?' I said looking round.

'Try to step on the drier bits. We'll walk around the edge.'

Droplets of moisture clung to my eyebrows and hair. Each year 10,000 millimetres of rain fall here.

'There are only twenty bogs like this in the Haleakala area,' Art told me, 'and as those on each island contain different plants and animals, these are probably the only ones like this in the world, so we figured it was worth hanging on to them. The clay silt sinks with all this rain; the nutrients follow, forming a hard rust pan about half a metre down which holds the water in. The colonizers have had to adapt to the boggy conditions. Take a look at this one.' He knelt down and held a small spiky green plant. 'This is a greensword, related to the silverswords you get down in the big crater. The same group of plants has adapted to everything from desert conditions to bogs. Incredible. And this is a geranium. It's more advanced than most others, and has these furry leaves like a rabbit's ears. Stops it dying from UV in the harsh sunlight when the mist clears up here. It doesn't grow anywhere else in the world. I named it myself.'

We moved to the bushes at the edge of the bog. He pointed to a tree about twice his height and as thick as his arm. The trunk was smooth, and had no branches at all. A rosette of leaves sprouted from the top. It was a lobelia. Most of us know lobelias as pretty little plants in gardens, but those that managed to colonize Hawaii, probably millions of years ago from America, underwent an evolutionary change that is often found on the remote Pacific islands: it is known as 'arborescence' which means growing like a tree. One of the best examples is *Euphorbia*, a common garden weed which forms low mats over paths and stones. The Hawaiians call it *akoko*. Coastal *akoko* are similar to these weeds and have floating seeds, which probably accounts for their arrival on beaches. Their leaves are still succulent, retaining moisture in their salty environment. Inland *Euphorbia celastroides* grows in dry, scrubby areas, and has picked itself up off the floor a bit, some varieties forming large shrubs. The plant grows leaves to match the moisture available at each site. In the dry Ulupalakua lava fields on Maui, it reaches the height of a small tree, while in wet boggy areas on Kauai, *Euphorbia remyi* has developed into a climber to reach sunlight in the forests there. *Euphorbia rockii* grows only in the wettest forests. It reaches a height of seven metres or more and has huge, leathery, dark-green leaves typical of rainforest trees.

The reverse process can also occur. In bogs like the one I was standing in, small violets grow with stems a few centimetres high, resembling those in the Americas. In the upper dry forest a giant variety has evolved, 25 metres tall. Plantains, which spread their tough leaves over pavements and gravel paths in our towns, have in Hawaii grown to monsters almost a metre tall, with enormous leaves as long as a man's arm. When these plants arrived, either floating or stuck to the wings of birds, they found lava beds ideally suited to their weedy existence.

The warm rainy interiors of the islands offered ideal conditions for forest, but there were few trees there other than those like the *ohi'a* with its lightweight seeds; most trees had seeds poorly adapted for long-distance dispersal. Evolution has played its hand to fill the numerous niches with plants that would never normally have had that chance. The plants here are the stuff botanists' dreams are made of.

Around one edge of the bog there was a tangle of wires and the beginnings of a strong wire-mesh fence. Escaped mammals – pigs and goats – are a major problem in Hawaii. The pigs which were originally imported by the Polynesians have interbred with stock from Europe, live wild in huge numbers in the forest, and are a favourite with hunters. They lose all resemblance to domestic pigs after a few generations in the wild, becoming big and mean. The boar's sharp tusks can deliver dangerously savage blows. The rooting behaviour of these pigs destroys tree seedlings in forests which evolved without having to cope with this predation. Pigs also love bogs and zero in on favourite plants, destroying many others with their trotters. So far four of the bogs at Haleakala have been fenced in a last attempt to save the rare plants inside. In these four, the proportion of native plants has increased from 6 to 90 per cent. The Parks Service removed 20,000 wild goats from the Haleakala National Park before they gave up and decided to fence that too. The damage the goats caused to the native vegetation was enormous, here and elsewhere. Sixty per cent of the Park is fenced off so far. It is a massive task, since the areas to be fenced include thick forest as well as bare cliffs. It has taken ten years to cover fifty kilometres. To protect Kipahulu alone cost a million dollars.

Kipahulu was our next port of call. The helicopter was manoeuvred through a small gap in the canopy, its rotors seeming almost to touch the swaying ohi'a leaves as we landed.

There are just over fifty species and subspecies of native land birds inhabiting the Hawaiian islands today, including a goose, a crow, an owl, a hawk, two thrushes, an old-world warbler, an old-world flycatcher, and a family of largely nectar-feeding birds, related to finches but hardly resembling them, called the Drepanidinae. All these birds have evolved here from perhaps as few as fifteen original colonizing species. In the Drepanidinae, or honey creepers, an evolutionary explosion has occurred. Forty-seven of them seem to have evolved from just one finch. One of the main reasons for protecting Kipahulu is that it contains some of the rarest of them.

Like Darwin's finches, the Hawaiian honey creepers became specialists and diversified into many forms to take advantage of the plants they found and in most cases the abundant nectar these provided. This 'adaptive radiation' is another feature of island evolution, and Hawaii's honey creepers provide one of the finest examples of it anywhere. I marvelled at the i'iwi, the brightest bird in the islands, scarlet with black streaks to its wings, for its ability to probe flowers with a long downward-curving bill that perfectly fits the blossoms of ohi'a trees and giant lobelias on which it depends for food. A smaller red bird with an engaging white fluffy behind, the apapane, is more common. It too has specialized to take nectar from the abundant ohi'a trees. Tens of thousands of apapanes will fly into areas where they are blooming. The palila, by contrast, has retained a stubby finch-like bill which enables it to feed on the seed pods of the mamane tree. The akiapolaau is bright yellow and confined to the 'Big Island'. Its bill is very long and curves downwards, but the lower portion is short like a dagger. With its lower bill it furiously pounds into the bark of koa trees like a woodpecker, winkling out grubs with the pointed upper half. Most extraordinary of all, perhaps, is the laysan finch, Telespyza cantans, the only surviving finch on a small atoll far to the west in the Hawaiian chain, and probably the one most closely related to the original colonizing species. It has survived by retaining its tough beak to break into sea birds' eggs. Those eggs which are too tough the bird shoves around the rocks with its feet until they crack. Recently it has been discovered to suck blood like a vampire. It leaps on to the enormous wings of the albatrosses with which it shares the islands and pecks away until it draws blood, then dips its sharp stubby beak into the wound to drink. The albatrosses seem to tolerate this annoying behaviour.

As we trudged through the Kipahulu forests, thick with ferns, moss and other epiphytic plants, I was looking out for one bird in particular, Bishop's o'o. It is a large handsome black bird with a long tail and bright yellow tufts of feathers behind its ears and under its wings and tail. It has some of the most beautiful calls in the Pacific and was thought to be extinct for eighty years until one was rediscovered in the upper Hana forest in 1981. The o'os of Oahu and Hawaii are now extinct, and just a few pairs cling to survival on Kauai. Kipahulu is also a stronghold for one of the strangest-looking of the drepanids, the crested honey-creeper, which has a large tuft of spiky feathers on its forehead, perhaps designed to gather orange pollen from the ohi'a flowers on which it feeds. It survives only on Maui. Here also to be found are the Maui parrotbills, small parrot-like birds with stubby hooked beaks.

After a good walk searching in the forest, I came away disappointed. The rarity of the Bishop's o'o was all too apparent.

Today the course of evolution does not move alone through barely perceptible selective pressures, gently massaging the twisted helix of a genetic code over great lengths of time. Now there is a different, much more rapid pressure which is causing irreversible change more rapidly than ever before. The hand of man is on the islands and his grip is unrelenting. Fully 27 per cent of all the endangered species in the United States, of which these islands form but a tiny part, are here. The forces of nature's creative evolution and man's rabid bite of extinction are combined here as nowhere else.

In 1982 John Simpson wrote of Hawaii's birds in the *Nature Conservancy News*, organ of one of America's premier nature conservation organizations:

Of the original 70 forms known to have existed when Captain
Cook arrived in 1798, 23 have become extinct; 30 are classified
as endangered. Currently only sixteen are represented by viable
populations. Perhaps the 58 kinds of forest birds have suffered
most: 21 of the 23 extinct species or subspecies inhabited
Hawaii's forests; another 21 are endangered. Almost *half* the
birds listed by the US Fish and Wildlife Service as threatened or
endangered in the United States are endemic to Hawaii. In fact
more birds are threatened with extinction on the Hawaiian
islands than in any other area of the world of a similar size.

Since then things have not improved. The Hawaiian crow,
known as the *alala* to Hawaiians and considered an *aumakua*
or guardian spirit bird, is now virtually extinct in the wild, and
numerous other species amongst Hawaii's unique finches are
on the point of extinction.

One of the strongholds for forest birds in Hawaii is the Alakai
Swamp on Kauai. At approximately six million years old, Kauai
is the oldest of the eight high islands in the Hawaiian chain. Its
volcanic slopes have been cut by wind and rain over millennia,
so that it has a quite different character to the other islands. Its
jagged green skylines and cliffs are some of the most impressive
in the Pacific. I went there to see what remained of the birdlife
of the island, travelling by helicopter as there are no good roads
into the interior. Turning inland from the south coast we
entered Waimea Canyon, a huge gash in the mountains which
rivals the Grand Canyon of Colorado. The chopper weaved and
turned, riding like a roller coaster on the wind, skimming
knife-edged ridges which fall away to great chasms. A small
river was to be seen twisting through a fertile valley floor, but
no crops grow there now; the Hawaiians who used to live there
succumbed to European diseases like measles and smallpox
and have never returned. On either side the steepness of the
forest-covered slopes and drier brown mountainsides is unbelievable; they have been worn away by rainfall, which over
time has carved the canyon itself. The thirst of luxuriant
forests on north-eastern coasts in the Hawaiian islands is
quenched by moist trade winds, unlike that of forests of the
South Pacific where the trade winds blow from the south-east.

Passing north-west and over the ocean we saw the fluted
cliffs of the drier Na Pali coast, falling a thousand metres to the

sea. A thin line was just discernible through the forest, marking a route once used by ancient Hawaiians to reach their taro patches in the Kalalau valley and now used only by adventurous hikers. We turned inland again over the Alakai Swamp. Occasional sedge bogs glinted through the canopy of *ohi'a* trees. To reach the swamp would take a gruelling drive through the Waimea canyon to Koke'e State Park and then a long hike across the ridges to the forest. The jagged twin peaks of Waialeale and Kawaikini passed by, both rising to more than 1,500 metres, their volcanic crags and ridges now covered in forest. The helicopter seemed dwarfed by Waialeale's silent throat, once filled with magma, but as we entered the curved cliffs shaped like a giant horseshoe, the rotor blades seemed almost to touch the steep grey walls of stone. Spray covered the windshield. Veils of mist floated inside from numerous thin waterfalls which plunged downwards from the forest-covered cliffs above, almost drifting the 500 metres or more to the bottom. We were hovering inside the wettest place in the world, where an incredible fourteen metres of rain may fall in a year.

Picking up a car at the coast, I drove up the canyon to stay at the delightful log cabins of Koke'e Lodge. Early the next morning I walked into the misty forest trails with David Boynton, a local teacher and avid birdwatcher. Much of the forest was fairly open as a result of Hurricane Iwa, which struck Kauai on Thanksgiving eve 1982 and destroyed many of the coastal villages and upland trees. Now passion-flower vines, introduced to the island for their magnificent flowers, had taken over and scrambled over trees and shrubs, blocking out the light and hindering the regeneration of the forest. It is strange that a vine of such beauty has become a lethal pest in many parts of Hawaii. Crimson *i'iwis* have turned the flowers to their advantage, perching at their bases and breaking into them with their long beaks to suck nectar. On the path I sustained bloody scratches from thorny blackberry bushes which had also escaped into the forest. Being pricked by thorns was a new experience: almost no Pacific plants have them, having no need to protect themselves from browsing mammals. Introduced cows which grazed these forests before it was designated a nature reserve left the spiky blackberries, and these have prospered at the expense of the unprotected native plants which the cows devoured.

As we returned to the car park after a somewhat fruitless search, David pointed out an *ohi'a* tree in bloom close to some cars. In the space of a few minutes I saw more rare birds there than I had seen in the whole of our walk in the forest. As is so often the case, birds which are used to humans are tamer, and a car makes an effective hide. A small olive *amakihi* dropped in to probe the tree's red flowers, and a *akikiki* creeper poked around the bark. As bright as a canary, a tiny bird with the delightful name of *anianiau* perched on a huge pink hibiscus flower which hung upside-down and probed inside, looking like an avian version of His Master's Voice. None of these birds were to be found anywhere else in the world, yet few of the tourists passing that way appeared to notice that they were there.

Kauai is more fortunate than the other islands. All thirteen of its original forest birds still survive in the Alakai Swamp. To be an uncommon species is an enviable status in these islands, but six of these birds are making their last stand in the swamp. The Kauai *kama'o* thrush is down to 300 or less. In 1890 it was the island's most common bird. The Kauai *nukupu'u* is rarely seen beating its strange beak into the bark of *koa* trees, and the extraordinary *akialoa*, an olive-yellow bird with a wicked curving bill a third as long as its body, has not been seen since 1973. Saddest of all is the Kauai *o'o*. Like the Bishop's *o'o* on Maui, this handsome black bird, a spectacular relative of the South Pacific honey-eaters which I had seen in Fiji and Samoa, was once sought by Hawaiians for its yellow thigh feathers so that they could add the finishing touches to the magnificent red-feathered cloaks and helmets that were the mark of Hawaiian royalty. In the early 1890s the birds were common and the forest was filled with their beautiful song. By 1973 a mere forty birds survived. Now in the mornings in the Alakai, a single, haunting and persistent call is occasionally heard. It belongs to one lonely male, the last of the *o'os*, who sings each day for a mate. He sings a requiem for his species, for no female now exists.

Reasons for this catastrophic decline in Hawaii's forest birds are hard to pin down. Certainly forest clearance has taken its toll, but the chance introduction of the southern *Culex* mosquito in the mid 1800s may have been more destructive, for it is the vector of avian malaria. This has swept the lowlands, but mosquitoes find the forest too cold above 900 metres and so the birds above that altitude have fared better. The northern *Culex* mosquitoes are more hardy, but have yet to arrive. When they do, and they almost certainly will – more than a dozen exotic insect species are introduced to Hawaii each year, despite every effort to fumigate air and sea traffic to the islands – it could spell doom for the varieties of birds surviving in the uplands.

Some native birds do persist in lowland areas. Perhaps they are resistant. It is significant that the most common of the Hawaiian finches, the *apapane* and *amakihi*, show the highest resistance to malaria in tests. More than 170 foreign bird species have been introduced to Hawaii, bringing new diseases and competing for food and nest sites. Invading exotic plants such as the banana poka, an ornamental vine, and the strawberry guava, are spread rapidly by pigs. They crowd trees and shade the forest understorey, preventing seedlings of food plants from becoming established. The roof rat and the mongoose have also played their part, but thankfully the latter has never established itself on Kauai. A crate containing some for import was dumped into the sea by a forward-thinking seaman.

Given the depressing outlook for the birds of Hawaii, it is pleasing to tell of a relative success story. A last remnant of the eight or so species of goose that once existed in the islands still survives there. It is called the nene. There may have been as many as 25,000 nenes in Hawaii when Cook landed. The Hawaiian goose or *Nesochen sandvicensis* seems to be related to the Canada goose, and probably colonized the islands from North America. It is a handsome bird with a black head and long buff neck fluted with black stripes. Water has played so small a part in its life for so long that the webbing has been partially lost from its feet. Like many island birds, it was remarkably tame. This was its undoing. Unscrupulous hunters shot it for 'sport'; its eggs and goslings were attacked by pigs, mongooses, feral dogs and cats, all of which had been introduced to Hawaii over the centuries. By 1949, when conservationists decided on a last-ditch attempt to save the birds, there were only fifty left. A special breeding programme was desperately needed, but almost nothing was known about the bird's breeding habits in the wild.

To attempt to breed young from the few remaining wild adults could have been disastrous, but fortunately a Hawaiian rancher, Herbert Shipman, had been rearing his own captive flock for thirty years in his garden near Hilo, the country seat of the Big Island. Prompted by calls to save the birds from Sir Peter Scott of the Wildfowl Trust at Slimbridge in England, a breeding programme was started. Pairs were loaned from Shipman's flock; several went to Slimbridge and others settled into a special Government centre established at Pohakuloa, 2,000 metres up on the saddle in Hawaii. In December 1949, the first two goslings were hatched, but little over one in five eggs were fertile. It seemed the Shipman birds were too inbred to be strongly fertile. The nene's prospects were not looking good.

When some birds were captured from the wild and added to the programme's flock, egg fertility nearly tripled. In 1960 the first pen-reared nene was released into the wild and joined the small flock flying free on the island. Since then more than 1,000 birds have been released into special sanctuary areas, over 200 from captive-bred colonies at the Wildfowl Trust. Now most birds tend to live in the particularly harsh surroundings of upland lava fields on the Big Island, nibbling on grasses and berries growing on the slopes of Mauna Loa and Hualalai at elevations between 1,700 and 2,500 metres. A second flock has been established in the Haleakala crater on Maui with birds from the Wildfowl Trust. In earlier times some of the birds used to migrate down to the warmer lowlands to lay their eggs but now none do this.

They are only just holding on in the wild. For some reason the reintroduced birds have not multiplied as expected. Perhaps they were originally reduced to such small numbers by alien predators that inbreeding resulted in lower fertility. Perhaps the captive-bred birds do not breed successfully with the wild ones. There is still a lack of information on what happens to the birds after they are released. Hawaii's rugged interior makes such information hard to obtain. One cynical biologist told me that the State captive-breeding programme merely served to feed the mongooses on the island. It may be that the nene cannot keep up with the predations that still

Paradise in the Balance

The pace of change in this beautiful ocean world has begun to quicken. New colonists bearing the flags of consumerism, tourism and development threaten the once-pristine atolls and green volcanic islands.

1 **Exploited islands**: Guano from the millions of seabirds which once roosted here has made the Nauruans rich, but their island is now in ruins.

2 **Giants of industry**: An immigrant Kiribati worker on his way to fish in the shadow of the massive phosphate-loading gantries on Nauru's coral shore.

3 **Bird Island, Tetiaroa atoll**: Here I was able to watch seabirds breeding in relative peace. Only a limited number of people may visit the island, which is owned by Marlon Brando.

4 **'Gizo hawking service'**: The gentle pace of life in this Solomon Island town is still as alluring as ever.

5 **Festival store, Papeete**: The baubles of Western civilization are irresistible to almost every culture.

6 **Fish hanging at a Tahitian roadside stall**: The coral reefs once seemed inexhaustible, but polluted lagoons are unable to supply the needs of growing populations.

7 **Solomon Island rituals**: The mix of the old and the new is all too apparent as these dancers celebrate Prince Charles's birthday at Baramole village.

8 **Hermit crab**: A plastic jar makes a perfect home, but not all animals are so adaptable.

9 **Polynesian girl**: Flowers and brightly-coloured leaves abound to adorn the islanders in traditional style. This girl's features bear witness to the influence of European blood.

10 **Fishing for game**: Tourists bring much-needed foreign exchange, but often at great cost.

11 **Fruit-bat store, Guam**: Flying-fox fruit bats are imported into Guam from as far away as Samoa. This trade now threatens bats right across the South Pacific.

12 **Outline of bird petroglyph**: Memories carved in stone are all that remain of the natural history of Easter Island. As the population of one of the world's most remote islands expanded, stripping the land bare of trees, civilization gave way to chaos.

1

2

11

FISH
FRUIT BAT
CRAB

FOOD
STAMPS
WELCOME

4

3

10

9

8

5

6

7

occur, or that the ancient native grasses on which it evolved to feed no longer grow in sufficient abundance. As it is Hawaii's state bird, the Government is unlikely to allow it to vanish altogether. A few nenes have been naturalized on Kauai and there are now plans to reintroduce the bird to Lanai. Both islands originally had nenes and are for the moment free of mongooses. The species has had a reprieve, but its long-term future in the wild is far from assured.

If you are going to ruin a Hawaiian island it is probably best to do it on Oahu and probably at Waikiki. This beautiful bay was once an unremarkable marsh. It is now a magnificent semi-circle of windows, balconies and concrete condominiums looking out on to the most perfect warm sea. Watching the sun set here I could not help but admire the ability of man to turn such areas to his advantage; there can be no denying the pleasure this gives to millions each year. But it is sad that so few of them are aware of the riches the islands hold inland. No travel brochure ever tells them where to look.

The coming of man to these islands has had a profound effect. As I cast my mind back to the many islands I had visited, I wondered where it had all begun. The early Polynesians, the Melanesians and the Micronesians were all great seafaring peoples whose origins and great migrations to the islands where they now live are shrouded in mystery. What drove them to leave the lands in which they were born? What freak of nature or cultural upheaval prompted men to sail across thousands of kilometres of uncharted ocean to discover specks of land as far apart as Hawaii and New Zealand, Easter Island and Pohnpei?

Chapter Eight

THE COMING OF MAN

Huahine

> How shall we account for this Nation spreading itself so far over this vast ocean? We find them from New Zealand to the South, to these islands [Tahiti] to the North and from Easter Island to the Hebrides; how much further is not known.
>
> Captain Cook,
> *Voyage to the Pacific Ocean,* 1784

ith powerful arms the Tahitian grasped the shaft of his adze and swung the stone axe high over his head. Pausing only for a fraction of time to determine precisely where the blow would land, he brought it down, cutting an arc through the sun. When the blade landed splinters flew across the humid air, tinging it with the sweet smell of sap. The canoe would be months in the making. For the great voyages it would undertake, only the finest *tamanu* or *ati* wood, the strongest coconut-fibre twine and true love for the perfection of its smooth hull, polished with the skin of sharks, would ensure success – and perhaps the discovery of new lands. The lightweight wood of hibiscus trees would furnish the paddles, and ironwood that grew along the coast had already made the stem posts. When finished the canoe would have twin hulls 21 metres long, and a platform in the centre with a thatched shelter. It would support some sixty men, women and children, plus pigs, dogs, chickens and many other provisions. It was not always easy to predict how long the voyages would last, but with fish and turtles caught from the sea and a fair breeze, a great distance could be covered in a week. Chiefs would come from far and wide to commission their craft for the long voyages between islands. Huahine was the home of the master canoe-builders; only those of the nearby island Raiatea could rival them. The skill of centuries rested here.

The canoe-building site was near to the shore. Teams of

warriors would drag the huge double-hulled canoe on log rollers over the sand at the top of the beach, to the pass through the reef and out into the sea at high tide. Only war canoes were larger; they could exceed 35 metres and hold as many as 300 warriors ready for battle. At their launching it was customary to sacrifice slaves and throw their blood over the timbers, or place the live bodies of enemies between the rollers as the canoe was dragged to the sea.

The construction clearing was filled with the sound of falling adzes. There was a smell of molten breadfruit sap to be used for caulking the wood. Women were cutting the huge leaves of the *fara* palm into strips to weave them into sails and teasing fibres from coconut husks for twine. From time to time the chief would come by to inspect the work, carrying his carved staff and wearing a large whale's tooth around his neck to indicate his high rank. His hut was not far away, its roof newly thatched with grasses gathered from the hillsides inland. Each year they burned the hills; that way the grasses grew better, and it kept the forest back. Most had been cleared from the lowlands and valleys now. Further up the coast was the chiefly village of Maeva, a whole string of stately residences positioned around a pleasant inland lagoon. On the hillsides there were other houses and many *marae* in which to worship the gods. Before work had begun on this canoe, the carver's tools and those of all the other *pahi* canoe-builders had been dedicated to Taaroa, the father of all gods, at the *marae*. A great feast of whole roasted pigs had followed, and the high priest had made offerings to the god of beauty and good weather, Tane.

On the day of *fa'ainuraa i te vaa* the craftsman of the new canoe would sing a simple prayer song:

> If I sail my canoe
> Through the breaking waves,
> Let them pass under,
> Let my canoe pass over,
> Tane!

Small blue lorikeets screeched their high pitched 'schee-schee' between the tall palms which shaded the site. Above, white-tailed tropic birds wheeled in the sun. The first inkling that something was wrong was a strange roaring sound coming from the reef, almost like the rumbling of thunder. The men stopped their work and looked about; there were no clouds, no signs of a storm. Then suddenly there was a great cry of 'Are miti rahi' from the beach, the sound of children screaming in fear, and men and women came running through the low bushes on the shore.

A vast wall of water rushed up the beaches, smashing the fringing ring of palms like matchwood, and as the canoe-builders turned to run they could see it clutching over them. The Chief's hut and all the others in the village nearby were swept before the advancing sea, carried inland in a mass of broken tree limbs, soil, sand, and floating debris. When the waves receded and drained back into the sea, the canoe-builders' site had vanished beneath a wasteland of mud as if it had never been.

Out in the deep ocean, tidal waves move at incredible speeds; 400 kilometres an hour or more is not unusual, but because of the great volume of water there they result only in a series of quite shallow swells. They are caused by the ocean floor collapsing after a volcanic eruption or by the Pacific plate moving into one of the deep trenches which surround this great ocean and causing an earthquake. Once a tidal wave approaches shallow seas, the pressure wave slows as it drags across the bottom and is deflected up towards the surface, sometimes building to an enormous height. Then it will race across reef flats and engulf whole coastlines with millions of tons of water. Following Krakatoa's eruption in 1883, 32,000 people lost their lives in tsunamis.

In 1972 a hotel construction crew began dredging a large pit near the coast on Huahine, one of the Leeward Islands in the Society group, 180 kilometres from Tahiti, capital of French Polynesia. The site was half a kilometre from the picturesque village of Fare, the capital of Huahine, situated on the west coast beside the Ava Mo'a pass which provides access from the harbour through the reef to the sea. With the first few scoops a number of curious wooden and bone objects were found which appeared to have been made by man. Fortunately, Yosihiko Sinoto, one of the most accomplished archaeologists in the Pacific, happened to be reconstructing an ancient Tahitian

meeting house at the now ruined village of Maeva further up
the coast. The hotel architect, Richard Soupene, invited him to
take a look at the finds. When Yoshi, as he is known, first
saw one bone artefact, he was stunned. It was a *patu* hand
club, shaped like a short paddle and fashioned from whale
bone. Nothing like it had ever been found before outside New
Zealand.

Here at last was evidence which corroborated ancient oral
legends of the Tahitians; evidence that New Zealand was
colonized by Polynesians who had sailed from the Leeward
Islands many years before. The site uncovered at the hotel was
clearly of enormous significance. Extensive excavations followed in 1974 and 1975, during which numerous artefacts were
unearthed dating back a thousand years. It also became clear
that much later the site had been engulfed with mud and sand,
perhaps from a large tidal wave, and an arc of deposited stones
revealed the direction from which it had come. There were
basalt adzes, stone scrapers and choppers, chisels made from
Terebra shells and turtle bone, and a large collection of shell
scrapers and graters, as well as intricately carved fishhooks of
mother-of-pearl. A wooden bow indicated that the aristocratic
sport of archery was practised. There were whale-tooth
pendants as worn by those of high rank, and even a chief's staff.

In 1977 there was another amazing discovery. Further dredging turned up more wooden artefacts including the boom of an
outrigger. Sinoto begged for dredging of the site to be halted
until further funds could be raised from the National Geographic Society. The hotel agreed. New archaeological work
began and discovered a large steering paddle, smaller canoe
paddles, an unfinished canoe-bailer, and two large wooden
planks seven metres long. The first remains of a Polynesian
long-distance voyaging canoe had been unearthed. From the
positions of the finds it appeared that they had been deposited
in the backwash of the tidal wave.

Mud had protected the wooden artefacts from decay. They
can be seen today in the Bishop Museum in Honolulu, and bear
witness to an important canoe-building site. In Cook's day the
people of Huahine and Raiatea were well known as master
builders of long-distance voyaging canoes. The discovery of
storage houses, probably for yams, at the site indicated a

prosperous people, producing more than their daily needs so that some could specialize in skills other than farming. Adze-making workshops and shell-scraping sites were also found. A community of perhaps 200 people had lived here, and the large number of beautifully carved pearl-shell products suggests that they were probably involved in trading these goods to other areas. The site revealed a people with complex social organization, a society well equipped for life on the islands and with the necessary tools to voyage great distances between them. Polynesian navigators were crossing the Pacific long before Europeans even thought of exploration. While the Greeks were felling the walls of Troy, Polynesians had already reached Fiji in their double-hulled canoes. Their voyages were far greater even than those of the Vikings.

Who were these men? Were they North American Indians or from the ancient mountain civilizations of the Andes as Thor Heyerdahl would have us think? Did they come from Southeast Asia, island-hopping their way east as most other archaeologists and anthropologists now believe? Today we are closer to a final answer than ever before, due to great advances in three areas: linguistics, blood protein analysis, and archaeology.

By examining the use of similar words and sounds in different island groups it is possible to see common ground between peoples widely separated by the sea, which provides clues as to the historical links between them. James King, Second Lieutenant aboard the *Resolution* during Cook's final voyage wrote of the Polynesians:

> It cannot but strike the imagination, the immense space through which this nation has spread, the extent of its limits exceed all Europe, and is nearly equal to Africa, stretching in breadth from A'toui [Kauai] to New Zealand . . . and in length from Easter Island to the Friendly Isles [Tonga] . . . All the isles in the intermediate space are by their affinity or sameness in speech to be reckoned as forming one people.

The word 'eye' in Tahitian is *mata*, in Hawaiian *maka*, in Maori *mata*. Likewise 'person' is, in the same order, *ta'ata*, *kanaka*, and *tagata*. In addition many words are borrowed from one language group by another, dependent upon the contact that exists between them. In the same way as *le weekend* is now part of French, the Fijian word for pig, *puaka*, was borrowed from the Tongans, who regularly used it to describe their feasts when they had conquered parts of eastern Fiji. Tongan loan words can even be found in kava and yam festivals as far away as Pohnpei in Micronesia. It is important for the linguist to distinguish between those words that are indigenous and those which are merely loaned in establishing the true language of any island group. Language, of course, evolves with time. Its changes are a valuable tool for the detective. By careful reconstruction of ancient forms of speech it is possible to trace the proto-languages which gave rise to those spoken today and this has been enlightening in tracing the origins of Pacific peoples.

It seems that a proto-Polynesian language, with root words common to most areas, links all the islands of the Polynesian triangle ranging from Hawaii to Easter Island to New Zealand. But there are many other small island communities outside the triangle which also share this language: the so-called Polynesian Outliers. These have been discovered in Melanesia and in Micronesia from the Loyalty Islands to the Carolines. Because this language shows remarkable similarity across its range, it seems that the original community which spoke it dispersed in relatively recent times.

Polynesian can itself be traced back further to a much more widespread language base known as Austronesian. With over 500 daughter languages, this is the largest well-established language family in the world, linking places as far apart as Madagascar, Indonesia, Taiwan, the Philippines, parts of Indo-China and the Malay Peninsula, right across the Pacific to Eastern Polynesia. Only in some of Eastern Indonesia, mainland New Guinea, and parts of the Bismarck Archipelago, the Solomons and the Santa Cruz Islands are non-Austronesian languages spoken. Though they are not all related these are collectively known as Papuan.

All this evidence together suggests a gradual movement of Austronesian-speaking peoples from Indonesia through the Melanesian Islands and out into the Eastern Pacific. A proto-Oceanic language seems to have developed in New Guinea and parts of the Bismarck Archipelago, with proto-Eastern Oceanic on Vanuatu, parts of Micronesia and Rotuma. Proto-Central Pacific evolved in the Fijian Islands, giving rise to proto-Polynesian.

Shared proteins in blood, and in particular blood antigens, substances which stimulate the production of antibodies, also point to relationships between peoples. The incidence of genetic diseases passed from one island group to another, such as the inherited blood disorder beta-thallasaemia, allows biochemists and geneticists to trace the origins of peoples. All of these, along with most of the animal and plant species inhabiting the islands, point to a Western origin for Pacific Islanders. It is, however, to the historian and archaeologist that we must turn for the most compelling evidence of the origins of man in the Pacific, and here there has been much controversy.

Sixteenth-century explorers were not impressed by Polynesian sailing canoes, believing them quite incapable of bringing the islanders to their islands. The belief in a great Southern Continent as the source of Pacific peoples continued through generations of European explorers.

Forty thousand years ago the first hunter-gatherers were walking the hills of New Guinea, then joined to Australia by land due to a lowering of the sea. In New Guinea's highlands some of the world's earliest agriculture later developed, but these primitive peoples seem to have lacked either the skill or the desire to meet the challenge of the Pacific. Some could manage short distances between islands floating on logs, and they may have owned simple rafts or canoes, but the colonization of the Pacific had to wait for three innovations; these were brought by a different people who had developed them in the islands of Indonesia, the early Austronesians. First they had perfected the domestication of certain food plants they had discovered in their rainforests; secondly they had improved ways of capturing food from their reefs and lagoons, and thirdly they knew how to design ocean-going crafts capable of long-distance voyages. These people were the first humans to set foot on the Pacific's enchanted islands. In their passing they left many scattered signs; footprints that only archaeologists can discern of shell, bone, charcoal and pottery; the last of these has been crucial in uncovering the coming of man to the Pacific.

A remarkably uniform red earthenware known as lapita,

with characteristic stamped designs, has been most important. Its simple appearance belies its great age; the earliest found example dates from about 1600 BC. The exact origins of the lapita culture remain a mystery but its pots are scattered across the Pacific from New Britain and the northern coast of New Guinea to the great island of New Caledonia, across the island chains of the Solomons and Vanuatu. Its people must have sailed the 850 kilometres of ocean to Fiji, and lastly to Tonga and Samoa where their pots are also found. Here, at the gateway to Polynesia, the lapita people apparently lost their pots, for they are found no further east. It may be that the clay needed to make them was in short supply, or that a preferred form of cooking, such as the underground oven in common use today or the use of hot stones, made them obsolete. By the time Christ was born only simple pots were being made and by AD 300 their manufacture had ceased completely. Distance made regular two-way communication with this people's homeland to the west virtually impossible; isolated in the great islands of Fiji and Samoa and the atolls of Tonga a new culture with a new language began to develop. New tools such as basalt adzes, worked from the new volcanic rocks they found, enabled them to fashion finer canoes, and with that came a new mastery of the sea. In Fiji and Samoa over 2,000 years ago the first Polynesians were being born, and the final colonization of Oceania could begin.

It has long been believed that the Polynesians sailed directly into the Pacific from some region outside it. This now seems unlikely. The ancestors of the Polynesians were the people of the lapita culture whose lost homeland was perhaps in the Bismarck Archipelago to the north of New Guinea, but the evolution of their culture into that of the Polynesians can now clearly be seen in the archaeology of Tonga and Samoa over the last 3,000 years. Crucial to further migration was the development of sailing craft capable of sailing upwind and so more effectively heading into the trades which predominantly blew against them from the east.

Did the Polynesians embark on their migratory journeys on purpose or were the islands merely reached by accident? Computer simulations of random drift voyages show that chance alone cannot account for the numerous colonizations that occurred. Archaeological discoveries also point to carefully planned expeditions, taking a wide range of plants and animals intended to be of use on arrival. The Polynesians had every intention of making land; it was almost as if they somehow knew that new territories lay ahead. What could have given them such remarkable confidence?

Perhaps one answer is their attitude to the sea. To most Westerners, the sea is traditionally an adversary to be feared. To Polynesians it is home, more so than the land. The sea is to them a natural highway, supplied with ample food if you know how to catch it, which they did. Never have I come across a people so at ease with the sea.

Another reason for the Polynesians' confidence was that they had become masters of navigation. They could read the stars and memorized star maps prior to voyages; they understood the patterns of waves caused by islands beyond horizons; they knew how to interpret drifting vegetation, flight patterns of seabirds and the warmth of currents, perhaps even gaining clues from the kinds of fish they caught. Few people have such knowledge now. Only on Satawal in Micronesia is the wayfinding art still practised by one or two ageing men.

There are times of the year when the trade winds do not blow adversely. The wind pattern reverses for a week or ten days and at these times the early voyagers could set out in their canoes knowing that the prevailing winds would soon return and bring them safely home again if no land had been found. Westerly winds tend to blow for about a third of the year in Samoa and a little less than a quarter of the year in Tahiti but occasionally they will blow for much longer, such as during the El Niño phenomenon. These intermittent westerly winds could have been responsible for carrying the first canoes to the Marquesas, from which Hawaii and Easter Island may have been reached. The Polynesians were provided with another clue to assist them on their journeys, the importance of which I believe has been underestimated: the routes of migrating birds.

All regular seafarers note the passage of birds. They are company on a lonely ocean as well as effortlessly beautiful to watch. It may be that, like the Vikings, migrating Pacific peoples carried birds to release at sea in the knowledge that if they did not return, land must be nearby. Two birds in particular could have played a vital part in the discovery of Hawaii, New Zealand and perhaps other Pacific islands: the Pacific golden plover *Pluvialis dominica* and the long-tailed cuckoo. The former species migrates 9,000 kilometres each year from the tundra shores of the Russian and North American Arctic to Hawaii and then south to the Marquesas and the Society Islands. Each spring as the birds flew north to their Arctic nesting grounds, resplendent in their breeding plumage of black spangled with gold, yellow and white, the Polynesians would have wondered where they were going. The long-tailed cuckoo migrates north from New Zealand to avoid the southern winter and reaches many of the South Pacific Islands including Tahiti and the Cook Islands, from which New Zealand is believed to have been colonized. Their annual migration routes, observed by fisherman far out to sea, may have been a vital factor in the Polynesians' choosing a direction in which to sail.

The movement of people across the Pacific came in a series of fits and starts. The origins of the Polynesians seem to lie in the genes, languages and material culture of the peoples of Southern China and Taiwan, who began to migrate southwards and eastwards in about 4000 BC. Perhaps as many as 45,000 years earlier the Australoid peoples had already occupied New Guinea, but they failed to move out to the islands in any numbers. The lapita culture developed in island Melanesia in around 1500 BC. The next 600 years saw these people migrate eastwards as far as Fiji, Tonga and Samoa. In about 200 BC, they set out from Samoa and found, after a journey of 1,600 kilometres, the sharp volcanic peaks and green forested valleys of the Marquesas.

They had to adapt to new conditions. Being further north, there were fewer coral reefs here, and a shortage of fish; they experimented with new techniques to catch them. A range of bone and shell fishing gear, not seen in Western Polynesia, remains as evidence of their struggle for survival in the Marquesas. New speech, new technologies, a new culture evolved. Here the Eastern Polynesians flourished for a further 500 years before three of the greatest feats of human navigation were undertaken, the ocean voyages to Hawaii, Easter Island and New Zealand. Hawaii was reached 1,700 years ago and 100 years later some canoes beached on the tall forbidding shores of Easter Island. The Society Islands had in the mean time already been settled, and from there, in about AD 800, the prodigious journeys to New Zealand were made. A thousand years ago the great colonizing voyages to Melanesia, Micronesia and most of the other Pacific Islands were complete, and the Polynesians were the most widespread people on earth.

If humans occupied New Guinea at least 40,000 years ago, already having crossed quite substantial water gaps to have done so, why does there appear to be no evidence of them in New Britain and the Solomons — a relatively short distance east — more than 4,000 years old? It seemed until very recently that man must have been landlocked in New Guinea for 36,000 years. Then in 1987 Chris Gosden from La Trobe University was excavating a cave in the centre of New Ireland overlooking the sea when he discovered evidence of man that was 32,000 years old. That so few early traces have been found in the West Pacific may be a reflection more of the lack of funds for archaeology than of man's true migration pattern in the Pacific. In the next decade it may well be discovered that man arrived in the Solomons, Vanuatu and even Fiji far earlier than is now thought. The reason no evidence has been found may be that it lies beneath the sea, as a result of the rise in the world's sea level since the last ice age.

Much has been made of the heroic Polynesian tradition of finding islands. The discoverers were elevated to almost god-like status. What was it that made them undertake their journeys at all? Overcrowding, war and banishment may have been reasons. Often it was impossible to return on pain of death, or the physical distances prevented it. Easter Island never appeared on any early Polynesian maps even though their canoes reached it; the colonizers were never able to return home to break the news. It may be that the distant ancestors of the Polynesians migrated because of a great flood. The earth began to emerge from the last ice age 18,000 years ago, and as the ice melted, the sea level rose and began to flood the lowlands which joined Australia and New Guinea, and those of the Sunda shelf which once joined the great islands of Indonesia. To move inland to avoid the encroaching sea would have resulted in contact with other fierce tribes already occupying the land. Lowland people were forced into a maritime culture capable of travelling and maintaining contact between islands and living off the products of the sea. A mystery remains. The sea level rose fastest between 12,000 and 5,000 years ago, but there is no evidence of man east of the Solomons more than 3,500 years old. The reason may be that the earliest landing points now lie hidden beneath the sea.

If the ancestors of the Polynesians had embarked on their journeys prior to the rise in the sea level, the prospect of the Pacific Islands would have been far less hospitable than it is today. The lowered sea level made them virtual fortresses. They were surrounded by sheer cliffs, the edges of coral reefs which had once descended beneath the ocean. Now dry and covered in scrubby vegetation, they offered few points at which to land save where the cliffs were cut by steep gorges carved by rivers draining from the mountain island. Perhaps the first settlers could have inhabited caves in the coral cliffs, then walked up the steep valleys and explored the dry reef flats, covered at the time with trees. These early sites would not be found today because they are submerged and no one has cared to look for them.

It is odd that a zoologist should point out such an inadequacy in our studies of prehistory, but John Gibbons was a man who moved freely between apparently unrelated subjects, forged links where none existed before, and dusted down those which had long been forgotten. Seeing how animals needed to island-hop across the Pacific, he saw no reason why the hugely increased number of islands when the sea level was lower could not have helped humans too.

Before he drowned, he had been poised to investigate this extraordinary theory, first put forward by himself and Fergus Clunie, Director of the Fiji Museum. Leaders in Pacific archaeology such as Professor John Green at the University of Auckland dismiss the idea, pointing out that the reduction in sea level did little significantly to reduce the water gaps between island groups, particularly to the west of Santa Catalina on the eastern tip of the Solomons. Still, the arguments for early settlement sites on coastlines now sunk beneath the ocean remain compelling. In Micronesia there have been reports of divers finding a cave bearing the marks of fire on its walls, and containing stones in a circle like a hearth, many metres beneath the sea. Off the north coast of Australia, underwater archaeologists have found the remains of primitive houses beneath what was once the dry land of the Sunda shelf. The oldest known sites at Vanua Levu in Fiji are opposite a wide shelf over which the first settlers may have moved inland as the sea encroached.

There is growing evidence that someone had reached parts of Melanesia before the lapita people 3,500 years ago. Waisted axes, similar to those discovered on the Huon peninsula in New Guinea, known to be more than 40,000 years old, have been found in Bougainville, Guadalcanal, and Santa Catalina. At Poha cave on Guadalcanal in the Solomons human artefacts have very recently been discovered beneath those of the lapita people. Santa Catalina is as far east in the South Pacific as a people with simple watercraft and moderate navigational skill could reach. Beyond, the sea formed a great divide for both man and nature. On Santa Catalina today there is a strange dance still performed called *mako mako*. Their faces decorated with frightening patterns and wearing high conical masks, their bodies smeared with reddish clay, the men of the island silently mime an ancient legend in which 'the men of the trees', perhaps a primitive jungle folk, turn in panic as 'the canoe people' arrive from the sea. Primitive humans had almost certainly reached these islands before the ancestors of the Polynesians arrived; the question is, who were they?

When the ancestral Polynesians entered the Pacific, they did not come empty-handed. The jungle fowl, ancestor of the chicken, originally wild in South-east Asia, is now wild throughout most inhabited islands of the Pacific. From their Asian homelands they also brought pigs; some still travel long distances strapped to the outriggers of canoes. Dogs would also be wandering the decks of Polynesian double canoes, not intended as companions, but as food. These small brown or black dogs were an important part of Polynesian society. Their teeth would be made into decorative anklets, their hair into fringes for cloaks. The adult dogs were slothful, not hunters, but vegetarians. Often they were penned with pigs and fed on *poi*, a pudding made from taro roots. Once fattened they would be cooked in hot rocks and fed to chiefs, or strangled as offerings to the gods. Most valued of all were the delicious young puppies, though those that had been suckled by women, a common practice, were usually spared the table. The original native dog has now vanished from the Pacific, absorbed into breeds brought in by Europeans. The Pacific Islands offered little for them to hunt, so they never escaped domestication to survive in the wild.

The influence of early human migrations in distributing animals around the Pacific was not inconsiderable. Possums and even tree kangaroos may have been deliberately introduced to the Bismarck Archipelago for food, as well as fruit bats to Tonga and the Cook Islands. Skinks and geckos travelled all over the Pacific in the thatching of shelters aboard canoes. Small snakes, snails, insects and other creatures reached the islands as stowaways concealed in bundles of stored taro roots and yams.

The arriving Polynesians quickly cleared coastal woodlands to grow taro, yams, coconuts, breadfruit, arrowroot, and sugar cane for eating. The screw pine was planted around their huts for posts, and its leaves were used for mats and sails. Bamboo served a host of purposes including the making of nose flutes and stamping pipes. *Lagenaria* vines would grow into gourds for water; ginger was brought in and now enlivens the forest with its beautiful red flowers. There were the seeds of candlenut trees, whose nuts would provide oil for lamps when mature, as well as *Cordia* seeds whose trees would offer shade, and hibiscus for flowers and paddles. The wild ancestors of all these plants grow in tropical South-east Asia, providing further evidence that the Polynesians came from here.

There is one staple of the Polynesian diet whose arrival remains an enigma: the sweet potato. Its origins lie in the high Andes and it was present – principally in Easter Island, Hawaii and New Zealand – well before Europeans could have introduced it. It may be that the seeds of the sweet potato dispersed naturally to the islands, rafting on coconuts or even assisted by birds, but the distances are immense. Another enthralling explanation is that the Polynesians may have made contact with South America. It could be that South American people, as Thor Heyerdhal believes, carried the sweet potato on their rafts to the islands, but the seafaring tradition of the Polynesians was such that their migrations eastwards need not have stopped at Easter Island. We may never know if the Polynesians reached the shores of Peru or Ecuador but if they did they would instantly have recognized the potato tuber as a root crop they could adapt to their needs and would almost certainly have returned with it, particularly to Easter Island where the cooler conditions were less favourable for the tropical crops they had originally brought with them. The geographer Robert Langdon has recently discovered early translations which reveal that manioc was also in use on the islands at the time of the first European arrivals. Similar explanations may account for its presence, but the controversy lingers.

I visited the ancient ruined village of Maeva, further up the coast from the canoe-builders' site, then walked up on to Mata'ire'a Hill, overlooking the lagoon. Hidden in the forest here were many more *marae*, older than those below. Sixteen of these ancient temple sites have been restored to fine terraces stretching up the hillside. Papaya trees and hibiscus scrub are encroaching on them now. The hillside is planted with vanilla. I scrambled over the massive roots growing out of the walls of Mare Mata'ire'a Rahi, once the most sacred spot in all Huahine. Warriors had once guarded it day and night; it was deserted now.

From Marae Tamata Uporu on Mata'ire'a Hill there is a magnificent view down the eastern coast of Huahine. Steep hills reach for the sea here. Once they were heavily forested, but scrubby areas and occasional gardens cover the slopes now. Below, I could see the creek through which the lagoon waters flowed to the Pacific. This island is a very different place from the one the first arrivals found. The rails and other ground-nesting birds have vanished; many of the magnificent pigeons and the beautiful blue lorikeets which once flew among the palms have also become extinct, except on a few isolated islands. Many Pacific islands of today, though they appear picturesque and still rich in cultural diversity, are a shadow of their former splendour in terms of both their human and their natural history. I was standing on temple stones laid by proud Polynesians full of hope. I wondered when things had started to go wrong.

For years it has been believed that the greatest era of change to the Pacific Island environment and its wildlife began with the arrival of the Europeans. The pest animals, virulent diseases and alien attitudes they brought certainly cut a swathe across the Pacific, but in the last few years it has become apparent that the process began much earlier. When the first Polynesians arrived they quickly imposed their cultural heritage on a yielding landscape. Forests in coastal areas were felled so that crops could be planted; soon the destruction crept into the valleys and, as populations expanded, even on to the steepest mountain slopes. The eye may be pleased by green mountains and valleys covered in low creeping *Dicranopteris* ferns and pink orchid blossoms mixed with patches of original forest, but these are the remnants of a ruined landscape cauterized each year by fire to provide fertile ash. Imported weeds quickly escaped and fared well on the scorched ground, defeating less vigorous endemic plants. Devoid of trees, the regularly burnt, thin volcanic soils lost goodness and slipped down slopes, spreading over river valleys and on to the reefs to create coastal plains. Many ring forts lie beneath the mud of the Rewa valley in Fiji, lost beneath this tide of ecological change in the mountains. The increasing land on the coast could not compensate for lost land in the mountains; growing populations found themselves pressed for fertile farmland. No longer were the islands evolving alone. The Polynesians had unintentionally embarked upon a gigantic ecological experiment, and its consequences were to be disastrous for the natural life-support systems of the islands, endangering the precious resources upon which the immigrants themselves depended.

Much can be learnt about Polynesian eating habits by looking in their kitchens. Rubbish pits which are centuries old can be informative. As the decades passed, the number, size and variety of shellfish on the menu declined, and there was a dramatic reduction in turtles. Certain birds became 'off', and the number of fish fell. Clearly the table was becoming barer. Most dramatic of all was the disappearance of bird life. In New Zealand so much forest was lost that it was originally thought to have been the result of climatic change after the Ice Age. Now it is known that the first Maoris caused the trees to vanish, burning the landscape to plant crops and to hunt the extraordinary flightless moa. At one time there were thirteen or more species of this fascinating bird, some twice the size of an ostrich, others comparable to a turkey. Within half a century of their arrival the Maori had dispatched them all, along with some twenty other species of flying bird including ducks, geese, an eagle and a crow. They also destroyed the North Island fur seal rookeries.

Evidence of extinctions caused by the Polynesians in the islands of the Pacific was first found at Barbers Point on Oahu in Hawaii in 1976. Yoshi Sinoto was excavating a sink hole for early human artefacts when he discovered some bird bones and sent them to the Smithsonian Institute in Washington DC for identification. The results were stunning; the bones belonged to numerous birds that no one knew existed before, including a large flightless goose and an ibis. Snails associated with human occupation, as well as the bones of skinks and geckos introduced by man, proved that the birds existed at the time of the arrival of the Polynesians but quickly vanished. In Hawaii it seems the Polynesians extinguished perhaps forty species of bird prior to the European explorers' ever setting foot on the land. These discoveries have not gone down well with native Hawaiians anxious to maintain the myth of the Polynesians as guardians of Paradise. It may be that the Polynesians were no better conservationists than modern Westerners – although their tools of destruction were much less effective than ours.

Since then more evidence has been uncovered at archaeo-logical sites on other islands. The megapodes, once wide-spread in parts of the Central Pacific, were almost certainly clubbed to death by the Polynesians or vanished through the loss of their habitat. Now only those on Niuafo'ou in the Tongan Islands survive.

In Samoa the catching of flying birds became a Chiefly sport. In the forest on Tutuila in American Samoa, I was easily fooled into thinking that a star-shaped mound on a hilltop was the ruin of a fort. Many are in fact ancient pigeon-catching grounds, scenes of great competition and skill. Chiefs would appoint catchers renowned for their art. Young pigeons taken from the nest were blinded with birds' claws and then trained to fly to the left or right from a perch while attached to a line of coconut

sennit. Adult pigeons were attracted to these expertly flown
decoys and, as they fluttered to the centre of the star mound,
catchers leapt from their hiding places, armed with huge sweep
nets on poles, and scooped them out of the air.

The Pacific Islanders did not always wish to capture animals
for food. As their cultures became more complex there was a
need to trade goods and to evolve a currency. In Papua New
Guinea huge *kina* shells still decorate the chests of Big Men to
signify their wealth. Sperm-whale teeth in Fiji assumed huge
value as *tambua*, the currency of favours between chiefs, of
political alliances, and of lives. Today in the Solomons shell
money is still manufactured on Malaita and is traded extensively throughout the islands, principally to purchase wives.
Round chips broken from shells similar to those of oysters are
drilled and strung on to lengths of twine, or today of nylon
fishing gut. Then they are sanded to necklaces of smooth
disc-like beads of white, red and black. Those made entirely
from the red rim of the shell have the greatest value, being the
most laborious to obtain. As in the case of the stone money of
Yap, value is attributable according to the time spent in preparation. In the Solomons enormous bundles of white shell
money-strings are used to buy a bride: five bundles if she is
mediocre, ten if she is a catch.

To Melanesians and Polynesians, red feathers were also of
great value for decoration and trade. The magnificent red musk
parrots I had seen in Fiji are also found in Tonga, but they did
not get there naturally; they were kidnapped by the Tongans.
At the time there was a vibrant trade in the parrots' red
feathers, which were used by Fijians to decorate the edges of
fine mats for Chiefly occasions. The Tongans also valued them,
and used to sail in their double-hulled canoes to Fiji in order to
obtain them. Some of the parrots were spirited away and
released in Tongatapu; the trade in which Fiji had a com-
manding role in the Pacific was thus undermined. Once they
were more widespread in the islands; they survive today only
on the ancient island of Eua. The use of red feathers for
decoration spread all the way from the Solomons north to
Hawaii, and as far east as Tahiti. Small parakeets with red
plumage such as Fiji's *kula* parrot were also popular and traded
through the islands. Today these parrots and parakeets have

become scarcer and dyed chicken feathers now decorate fine mats.

Further to the west of Fiji, in the Santa Cruz Islands, red feathers carried an even greater value, and were bound into a fascinating story of wealth, spirits, the exchange of wives, and prostitution. The cardinal honey-eater *Myzomela cardinalis* is widespread in the South Pacific, ranging from Samoa to the eastern Solomons; related species also occur in Micronesia. It is smaller than a sparrow but styled like an emperor; the male has black wings and tail, while the rest of its feathers are splashed with scarlet. An elegant curving black beak allows it to probe the smallest flower, draining it of nectar with a forked tongue as efficient as a drinking straw. Red-feather money is only made on Santa Cruz Island by a few specialists whose perceived knowledge of the correct taboos and easily offended spirits that guard the forest traditionally gives them the ex-clusive right to manufacture currency. First a bird-snarer must fashion small perches covered in sticky latex which he posi-tions in a suitable tree, attaching a nectar-rich flower which is hard to resist or a live bird as decoy. Concealing himself behind a blind of palm leaves, he chirps on a special whistle made from a tree bud, so attracting the males to the sticky perch and capturing them.

Most birds die once the red feathers have been plucked from them, but in Hawaii, where magnificent cloaks and helmets were also traditionally made from the red feathers of the abundant honey creepers as well as the highly-prized yellow feathers of much rarer honey-eaters, it was considered a great skill to remove them delicately and release the birds to grow a new set.

Menfolk of the Reef and Duff Islands in the eastern Solomons would traditionally sail south, trading their women to Santa Cruz for feather money, which they were themselves unable to make. The feathers of the cardinal honey-eater were bound into belts up to ten metres long, using the plumage of over 300 birds. In the past those women sold as concubines would fetch ten times as much as brides. The women themselves, of course, derived no benefit from the trade. Concubines lived in the men's meeting house, and the purchaser could purvey them as prostitutes, deriving a high income from their services.

Feather money is still used occasionally for trading pigs, or even canoes. Inflation is negligible, because its value declines with age; the colour of the feathers fades even if they are wrapped in leaves and placed near smoky fires, and moths and mould also take their toll. The infiltration of the Australian dollar has, however, undermined the value of feather money. Although there are still few fathers who will marry off their daughters for dollars, the cardinal honey-eater's song is now heard more often in the forest on Santa Cruz.

Chapter Nine

VANISHING EDENS

The Society Islands

> We are an island nation gone mad, behaving
> like a limitless continent in a world that has
> already turned into a crowded island.
>
> Gavan Daws, *Shoal of Time*, 1968

ne of the finest islands in the whole of the Pacific must be Moorea. The steepness of its jagged volcanic peaks, sometimes swathed in layers of mist, is a masterpiece of natural sculpture. Twice as old as Tahiti (quite visible twenty kilometres to the south-east), its semi-circle of peaks is all that remains of a large volcano that once stood 3,000 metres high. Moorea is triangular, about twelve kilometres across, and surrounded by a barrier reef which is broken by twelve passes; these correspond to the twelve valleys of the island. Lagoons of aquamarine encircle the island inside the reef, tinged with white as surf breaks on to the corals. On the north coast there are two beautiful bays; one of them is named after Captain Cook, who sailed into it while surveying the Society Islands, as they have come to be known. The bays are offset by sharp ridges covered in contrasting patches of pastel *Pteris* fernland, pineapple groves and dark tropical vegetation, which rise to an amphitheatre of mountain pinnacles.

To enter Cook's Bay in a boat dwarfed by the high mountains is breathtaking. Tall coconut-palm groves cover the coastal strip behind a few sandy beaches, and the sun plays on the red corrugated roofs of small neo-Polynesian villas between them, which are a memory of the colonial vanilla trade. At the far end of the bay Mouaputa soars to 800 metres, its summit pierced with a curious hole through which the sky can be seen. Tahitian legend has it that Pai, the favoured child of the gods, fired an arrow through the mountain from Tahiti to prevent its being towed away by Hiro, the god of thieves. The arrow woke all the sleeping cockerels who set up such a noise that Hiro took fright and fled.

The pace of life has quickened on Tahiti since the 1950s, but Moorea has remained a more peaceful place. There are still numerous palm-thatched huts along the shoreline; outrigger canoes lie dragged up on the beaches; fishing nets hanging on frames in the sun play with the drying wind. A small road winds around the island with tracks leading off it; these are lined with pale green *riri* leaves backed with banana palms, to act as property boundaries. Everywhere there is the smell of frangipani bushes, the splash of enormous pink hibiscus; yellow creepers vie with deep red poinsettias. Girls stroll by, their flowing dark hair decorated with bands of white *tiare* flowers and green leaves.

But Paradise is in demand and changing fast. Flights from Tahiti arrive in Moorea every fifteen minutes. Mooreans commute the short hop by air to Papeete to work in the hotels and offices of the capital, returning home again at night. Large numbers of luxury hotels have sprung up all around Moorea's coast, and the tranquillity of the coastal road has been shattered by mopeds and minibuses.

Early explorers were disappointed to find the Pacific without silver or gold, spices or silks, but in China they found a market for one product the islands had in abundance: sandalwood. This rather unimpressive shrub has a fragrance quite irresistible to Chinese nostrils. Joss sticks, aromatic tapers and incense were all fashioned from it, and its oils were made into medicines and perfumes. European traders sailed mainly from Australia to New Caledonia, the New Hebrides (Vanuatu), Fiji and Hawaii, where the trees grew abundantly. In 1800 the castaway Oliver Slater marshalled the natives under his power in Fiji at what is still called Sandalwood Bay. Paying with axes, whales' teeth, powder and weapons, he stripped the islands of their precious timber within fifteen years. He was slain soon afterwards, the victim of intertribal jealousies. Ten more years saw the end of sandalwood in the New Hebrides, and in Hawaii all the available trees were felled by 1825. Now the slow-growing tree has vanished from all but the least accessible parts of the islands.

The behaviour of sandalwood traders was often reprehensible and coloured the islanders' opinion of white people for ever after. Cecil Foiljambe, aboard HMS *Curaçao* in 1875, had this to say:

> I find the sandalwood traders have much to answer for. They and indeed most of the white men in the Islands are the very scum of England and the Americas. They are afraid to show their faces in civilized places and make the poor natives what they are. They try to cheat them, practising greater cruelties than the cannibals themselves are capable of. For instance, one captain of a sandalwood schooner the other day boasted that, having taken his cargo of sandalwood, and as he was sailing along the coast he shot down unoffensive natives as they stood on the beach for the charitable purpose of spoiling the trade for the next comers . . . Can you wonder after this that they detest white men and would kill and eat them when a chance occurs?

With the collapse of the sandalwood trade, the islands were looted for bêches-de-mer, again for the Chinese market; these would be traded for silk, tea, jewels and precious woods. Though trading continued in Fiji until 1850, the sea slugs proved resilient and are still to be found in abundance on Pacific reefs. The emergence of sugar plantations in Fiji and Queensland in the nineteenth century created a new traffic, this time in the natives themselves. The notorious Blackbirders sailed the Pacific looking for easy pickings, filling their ships with islanders at gunpoint and carrying them to a life of slavery on plantations far from their homelands. Pearls and turtle-shells were also plundered to be traded through Canton. Then the whalers came. America dominated commercial whaling through the latter part of the nineteenth century, using the Bay of Islands in New Zealand's North Island as a base from which to scour the South Seas, and Hawaii to search the north. The oil of the cachalot was held supreme and so the sperm whale began its precipitous decline in the Pacific. Today the great whaling fleets have vanished, but despite a worldwide moratorium, limited numbers of whales may still be hunted illegally each year by unscrupulous fishing companies.

Meanwhile, Victorian missionaries quickly spread the word

through the Pacific that traditional practices such as singing
and dancing, wearing clothes made from leaves and worship-
ping ancient values were wrong. Numerous timbered white
churches and full-length Mother Hubbard dresses still worn in
Vanuatu are relics of those austere times, after which the
islands were never the same again. For the missionaries to have
brought cannibalism to an end in many of the islands was a
magnificent achievement, but by destroying the islanders'

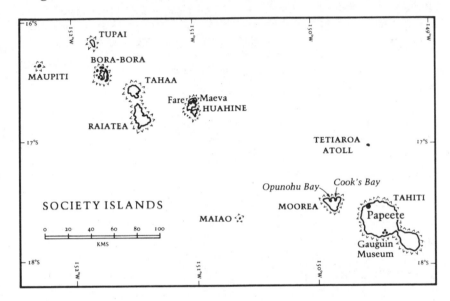

spiritual link to the land around them they destroyed their
system of *tapu*, the taboos which regulated their dealings with
each other and their use of the natural world. Once the power of
the chiefs and spiritual leaders had been undermined, the
exploitation of the land around them could proceed apace. To
Pacific peoples, land cannot be bought or sold, only used.
Unoccupied land is used as hunting ground, and so supports
the occupied land. Europeans brought the concept of land
ownership, and having instilled it used their power to wrest
land from the natives' grasp.

The first contact with the outside world began an inexorable
slide into doom, for even as the colonial powers grappled with
the spoils and annexed one island group after another, the
islanders themselves were dying. Tuberculosis, smallpox and

venereal disease, measles and whooping cough swept through the islands and decimated their populations. When Cook arrived in Tahiti, he estimated that there were 150,000 people inhabiting the green valleys and coasts. Just thirty years later the figure had dropped to 15,000, and by 1815 a census recorded a mere 8,000. In the Marquesas the population fell from 50,000 at the time of early European contact to just 2,094 by 1916. Elsewhere the story was the same. Without resistance to imported diseases, the Polynesian culture began to disintegrate. The very name of the Tahitian royal family, Pomare (*po* meaning night, and *mare* meaning cough), referred to the tuberculosis it suffered. In 1888 the last of the Pomare kings signed away his kingdom for a pension, and it has remained 'la Polynésie Française' ever since.

To the wildlife of the Pacific Islands these new developments initially made little difference. The Polynesians had wrought many changes to the landscape resulting in numerous extinctions, but the effects of European influence were to be greater. One of the most important of these was the conversion of the island's forests to agriculture on a massive and unprecedented scale. Huge copra plantations provided a trade in coconut oil which now feeds the western world with margarine. Almost a quarter of the world's copra originates in the islands. Native forests in the lowlands gave way to vast tracts of sugar-cane, pineapple, coffee and cocoa. The rainforests were felled to make way for farmland, and cattle graze where *ohi'a* and *koa* trees once stood on Hawaii. With their habitat destroyed and introduced disease weakening their bodies, many of the native wild species of the islands began to disappear as rapidly as the islanders themselves. This was also partly due to the other animals which were introduced to make sailors and traders feel more at home, or to be hunted as game animals, or to combat the increasingly troublesome pests which in the absence of the normal predators were beginning to ravage the plantations. Alien ants now creep over Hawaii's lava fields, attacking the bees which nest in them and so endangering the potential pollinators of plants. Wasps (*Vespula pennsylvanica*) are a new sight too and are causing havoc chasing and devouring native insects which have never seen them before. More than 800 exotic plants have now established themselves on Hawaii; in particular the strawberry guava (*Psidium cattleianum*), various fire-prone grasses and a passion-flower vine which often grows vigorously, choking native trees and shrubs.

The sheer number of introduced birds is staggering. Hawaii has received 162, more than all of North America. New Zealand now has 133 invaders from abroad, Australia 96. Tahiti and Fiji rank next with 56 and 25 apiece. Many introduced birds have become serious agricultural pests.

Most obvious of the avian interlopers is the noisy Indian mynah. Everywhere I went in the Pacific these small black birds with bright-orange beaks and legs seemed to flutter among the villages and farmland of the coastal fringe. Both the common mynah and the jungle mynah, which looks similar but has a tuft of bristles at the base of its beak, were introduced to the Pacific in an attempt to combat a plague of insect pests, particularly of stick insects. The birds themselves have now almost reached plague proportions, gathering in vast flocks on buildings at night to roost, squawking and chortling like hordes of hungry starlings. Resident bird-lovers will often say that introductions such as these have driven the native birds to take sanctuary in the mountains, and that that is why virtually none of the numerous attractive pigeons and songbirds which live in the forests of the Pacific's high islands are to be seen on the coasts today. A more likely reason is the conversion of the coastal plains from forest to modern agriculture and urban development. The native birds which had evolved on the islands were quite incapable of adapting to this alien habitat, but the introduced birds, many of which were imported because of their familiarity with homes and gardens – such as Fiji's bulbuls, which were brought in by indentured Indians – were ideally suited to the newly created coastal zone. The sad fact is that without regular sightings of their native birds and without a use for them in their culture, most islanders have now forgotten their names and few care that they exist. Persuading governments to take measures to save those that remain is therefore all the harder.

In the coastal coconut groves of Moorea I found hardly any native birds left at all. There were not many to begin with. I was now reaching the eastern limit to the colonizing potential of the animals, plants and people in whose footsteps I had trodden. From Tahiti a scatter of atolls known as the Tuamotu Archipelago fills a vast area of ocean to the north and east. There are few other high islands beyond the Tuamotus save the Marquesas, the desolate rock of Pitcairn 2,000 kilometres east, and Easter Island 1,900 kilometres further on. No doubt when the Polynesians first arrived there were bird species inhabiting Moorea which are now extinct and whose bones remain undiscovered. Since European times, the Polynesian ground dove with its purple-black back and white breast has vanished, as has the dark-blue lorikeet which once played among the palms of the coast.

From Cook's Bay I decided to drive round to Opunohu Bay and take the small road that led up into the mountains. What I was searching for was not a beautiful bird, or even a spectacular view, but a tiny delicate snail. On high islands in the Pacific basin there is a group of bean-sized tree snails which are not found anywhere else in the world. Their ancestors possibly reached the islands stuck to birds' feet many thousands of years ago, but since then they have evolved into numerous remarkable forms, each decorated with intricate lined patterns. Those of Mooréa belong to the genus known as *Partula* snails. There are seven species of them, originally separated by the valleys and ridges which dissect the island. For those not obsessed by tree snails, *Partula* is at first sight a boring little snail. It is mainly brown, though snail-lovers will sometimes describe it as pink. It does not travel very far and may spend its entire seventeen-year life in one tree, dining on tiny plants growing on the leaf surfaces, known as micro-epiphytes. To the trained geneticist, however, *Partula* is fascinating, because it is a key to understanding the origin of species.

These small snails are hermaphrodites, give birth to live young, have a relatively short generation time, do not escape enormous distances into the forest, and are easy to breed in a plastic lunch-box. Genetically, they are extraordinarily variable, a fact made clear by the patterns on their shells and even by their internal chemistry. When H. E. Crampton, the great collector of molluscs in the Pacific, made his classic study of *Partula* snails on Moorea in 1932, he was delighted. Here at last was the perfect place to get at the heart of the evolutionary process; nature had run an experiment in speciation on an isolated island. His hopes were not to be fulfilled, for today, after many thousands of years of harmless isolation, *Partula* has been eradicated from the wild as though it had never been.

The original reason for this decline was the predilection of the Governor of Reunion's mistress for Madagascan snail soup. In 1803, he imported some very large snails from that island which then escaped from his garden and became a pest, devouring the island's crops. The giant African land snail, as *Achatina* is now known, reached India in 1847 and by the 1930s began appearing in the South Pacific where it was introduced into French Polynesia in the belief that the Polynesians would like '*l'escargot*'. They did not, but the snails adored their fruit plantations and spread ankle-deep. It became imperative to find some means of controlling them.

Biological control is practised widely on islands. The idea is either to import a specific predatory animal or disease to destroy a pest, or to attract females to mate with an introduced sterile male of the same species rendering eggs infertile. In many cases this has worked well. Introduced parasitic insects have effectively controlled many pests in the Pacific. The rhinoceros beetle which plagued palms in Fiji was eventually controlled by an introduced virus after efforts with other beetles and assassin bugs had failed. Occasionally, however, the introduced predator or parasite behaves unexpectedly, and like a rogue missile targets creatures which it is not supposed to attack at all. That is precisely what happened with *Partula*.

By the early 1970s the Service Economique Rurale was desperate; Moorea farmers were being swamped by giant African land snails. Hearing of the success elsewhere in the Pacific of a small predatory snail from the south-eastern United States called *Euglandina rosea*, they introduced it to Moorea in 1977. The effect has been dramatic. *Euglandina* is a formidable predator. It sniffs out chemical scent trails left by its prey and follows them with the tenacity of a bloodhound at more than a snail's pace. *Euglandina* quickly discovered that the small *Partula* snails were an easier catch than the enormous African giants. Soon the defenceless *Partula* were being overrun as *Euglandina* spread through the forest at the rate of over a kilometre a year, devouring every small snail in its path. By

1983 the native mollusc had vanished from much of the island.

Wherever *Euglandina* was introduced to combat the African land snail, the pattern was the same. Colourful *Achatinella* snails endemic to Hawaii take six years to mature and only produce one or two young annually. Six hundred released *Euglandina* snails increased to 12,000 in just three years. Half of the forty-four species of *Achatinella* unique to Hawaii have been extinguished already. Once the process starts it is hard to stop. The extraordinary thing is that governments continue to recommend the introduction of the American predator snail despite these disastrous consequences. Local farmers believe it works, and have no interest in the welfare of their native snails.

In a desperate bid to save Moorea's *Partula* snails, scientists gathered specimens of as many species as they could find and posted them to breeding centres around the world. Alas, little snails fire few imaginations, but despite the lack of funds some were established in breeding colonies, and just in time. When I went to search for *Partula* among the rainforest leaves of Moorea, I could find none at all. A full survey of the island in July 1987 revealed that all Moorea's *Partula* snails are now extinct in the wild. Now I understood the significance of a small tank that sits in the reptile house of Gerald Durrell's zoo in Jersey, my native island. Inside are all that remain of the snails, from an island of equal size on the other side of the world. Few people visiting the Jersey Wildlife Preservation Trust notice them, but it is precisely this sort of quiet effort that must continue if we are to stem the loss of the genetic inheritance of our planet. One day it may be possible to reintroduce these delicate little snails and the evolutionary secrets they contain to their homeland.

From the hills inland I could look across to massive Mount Rotui, which divides Cook's Bay from Opunohu. The view in the evening light was magnificent even though there was little left of the natural forest which had once clothed the island. Hibiscus bushes dotted the slopes now, between flanks of planted *Casuarina* trees and fields of pineapples. The mountain's jagged outline, like a giant's green castle, was lit with golden sun from the west, with water surrounding it on three sides. Quite suddenly it was almost dark, and hawk moths appeared like small fluttering ghosts among the flowers, seeking nectar with their fine long tongues. Absorbed by my thoughts I was startled by eerie cries from the mountains behind me, a distant wailing sound, echoing among the dark crags above my head. The hair stood up on my skin. There was no one around. The cries increased in volume; it was as though the spirits of the mountains, thrilled with the dark night, had suddenly come alive. With relief I remembered: these were petrels, nocturnal seabirds, which nested in burrows high on mountaintops out of the reach of man and many of the savage stowaways he inadvertently brought with him.

From the first time that man journeyed into the Pacific he has been accompanied by animal stowaways, and some of these have waged war with the animals they found inhabiting the land. The lapita people brought a small rat with them from South-east Asia. They did not do so intentionally, but archaeologists have found that wherever they went in their canoes, the rat went too. The Polynesian rat *Rattus exulans* fed mainly on their stored crops and appears to have had little effect on the native creatures with which it shared the islands. Not so the black rat *Rattus rattus*, which began scurrying down the mooring lines of ships docking at new ports in the Pacific in the sixteenth and seventeenth centuries and has been doing so ever since.

The black rat was an excellent climber and took to eating native lizards and eggs from birds' nests in the trees. It was already established on Galapagos prior to Darwin's visit in 1835, and dispatched five species of rat found only on those islands. Its introduction to some oceanic islands has resulted in ornithological catastrophe. On Lord Howe Island off Australia's south-eastern coast, black rats destroyed 40 per cent of the indigenous species of forest birds within five years of their arrival in 1918. On Big South Cape Island, off the south-west coast of Stewart Island in New Zealand, the populations of eight species of landbird plummeted within three years.

For some reason this does not seem to have happened in the most tropical Pacific Islands, perhaps because the birds here were already well used to dealing with another predator of their eggs and chicks: land crabs. The enormous robber or coconut crab *Birgus latro* was once widespread on many atolls and islands of the Pacific. The Polynesian desire for them on the table has resulted in greatly reduced numbers but they can still be found creeping about in the forest far from the sea. The robber crab is a huge hermit crab – larger than a lobster – which has earned its name for its habit of climbing palm trees to snip down coconuts, which it then breaks with monstrous pincers to feast on the white meat inside. These and other land crabs are fond of creeping into birds' nests to dismember chicks, particularly those of seabirds. It may be that the danger of such a hideous end has made seabirds which share their islands with such crabs wary of all small furry creatures with designs on their young. I find this hard to believe, however, as I have read of rats on Ile du Lys in the western Indian Ocean launching a mass attack on thousands of sleeping noddy terns roosting in shrubs on the atoll's shore. Shortly after dark the rats began creeping up the branches, then leapt at the terns' throats, wrestling them to the ground and fighting them to the death. The slaughter continued until midnight, when the sated rats appeared to retire for the night, as did the two ornithologists who had chanced upon the event. The rats then started to harass *them*, nipping limbs and stealing food until the observers were forced to flee the island.

The brown rat which soon followed the black is less of a climber, so it had little effect on perching birds. Instead it invaded the new sugar plantations in such numbers that the natives refused to work them during harvest-time, when the cane was burnt and hordes of vicious rats ran amok attempting to avoid the flames. In an effort to control the rats, planters introduced one of the world's most efficient and ruthless killers to the Pacific Islands, the Indian mongoose. One of the best places to see it is the Thurston Botanical Gardens in Suva. There are always a few playing around the colourful borders, their inquisitive faces poking between the flowers, before they scurry on tiny legs like clockwork across the neatly mown lawns. The mongoose was brought to Fiji from Hawaii in 1873 and released on Vanua Levu and Viti Levu. The planters, having little knowledge of the natural world, had overlooked a simple fact all too apparent to anyone sitting in the botanical garden: the mongoose hunts by day. It therefore rarely met the nocturnal rats, which continued to flourish. Instead, it cast its expert eye on native birds in the bush and fell on them with relish. It has been responsible for the extinction of no less than seven species on the Fijian Islands alone, including the purple swamp-hen *Porphyrio porphyrio* and the banded rail *Rallus philippensis*. Its effects on the ground-nesting birds of Hawaii and everywhere else it has been introduced have been similar.

Throughout the Pacific, animals introduced by man have had a devastating effect on bird life. Feral cats, originating as pets, have played havoc in seabird-nesting grounds such as those on Christmas Island. The introduction of the great horned owl *Bubo virginianus* may have been responsible for the disappearance of the beautiful red-moustached fruit dove *Ptilinopus mercierii* from the Marquesas. It is rare, however, that an introduced animal causes wholesale slaughter of numerous different kinds of bird, driving some of them to complete extinction. It is more unusual still for the culprit to remain undetected. In the 1970s something began to decimate the bird population on the island of Guam in Micronesia, but nobody knew what it was.

The forests there are silent. The birds which once filled Guam with song have vanished within the last fifteen years. Once it was possible to take a stroll through marshes and woodlands and spot the handsome dark-blue head of Guam's unique flycatcher or hear the calls of the brindled white-eyes with their pale-green plumage and yellow breasts, busily foraging for insects. The rufous fantails no longer wiggle their pretty tails at passers-by; the kingfisher and the rail vanished from the wild the very year I went to see them. The rail was found nowhere else but on Guam. Only a few introduced species of bird now remain; the eighteen native birds have all but vanished. As recently as 1968 it was believed there were as many as 80,000 rails on Guam. What could possibly have caused such a catastrophic decline so quickly?

The alarm bells started ringing in the late 1970s when a census revealed a precipitous drop in the number of every kind of bird on the island, even in areas where human development had not taken place. In 1979 Guam requested that ten of its birds be placed on the US Endangered Species List, to obtain funds for their protection. No action was taken until 1984, by which time it was almost too late. In Guam I met Julie Savidge, a biologist from the University of Illinois who had come to study the problem in 1982.

'At first I thought it was some kind of bird disease that was sweeping the island. I checked parasites and bacteria, and stuck birds in cages where they could be bitten by disease-transmitting insects, but none of them caught anything. The next possibility was pesticides. I soon found that on islands using similar pesticides to those used on Guam there was no such dramatic decline. The same was true of hunting. I also discounted introduced animals. Rats seemed to be having no effect and monitor lizards were not affecting birds on other islands, though they will eat birds. By now I was getting pretty desperate. All around us the birds were disappearing, and I simply could find no reason for it.'

By 1983 the only place where the ten native forest species could be found was a small strip of mature forest at the island's northern tip. Then came the clue: native Guamanians said a snake was eating their birds.

'There is one tiny snake, blind, which burrows in the soil,' said Julie. I remembered John Gibbons's *Ogmodon*, to which this one is related. 'It's far too small to be a danger but there is one other, a brown tree snake accidentally introduced to Guam after the war.' It had to be *Boiga irregularis.* No snake had ever caused extinctions before but this snake had explosively increased its numbers over the same period as the birds had declined. Julie conducted exhaustive studies and concluded that the snake was indeed the culprit.

These reptiles have now become so numerous that they climb up telegraph wires and short Guam's electricity supply. Snake carcasses hang from power lines beside roads, grisly reminders of the peril lurking in the grass. There may be as many as 6,000 per square mile in certain parts of the island; the total population must be counted in millions. The brown tree snake also occurs in coastal Australia, New Guinea, and Sulawesi, but it seems larger and more voracious on Guam than anywhere else. The population seems to have got out of control, which is a classic result of an alien's introduction on to a tropical island without the biological brakes that hold it in check in its home. The snakes are poisonous and efficiently seek out their prey by smell, engulfing roosting parents, chicks and eggs; their mouths are enormous, capable of swallowing a chicken. The birds of Guam, unused to such capable predators, do not build their nests on inaccessible branch tips, nor have a language of alarm calls to warn their neighbours; they have paid the price for their evolutionary naivety. Even though Julie has fingered the snake, it seems unstoppable.

In an attempt to capture some of the few remaining Guam rails to breed in captivity, the US Fish and Wildlife Service took Koko on to the payroll. Koko was a tame rail which they wanted to release to lure wild rails from hiding in the last place they were known to exist, on Andersen Airforce Base. The bird immediately became entangled in red tape. Three old rusting bombers near to the rail habitat were to be blown up as part of the Salt II Treaty with Russia. The rails would have been frightened by this, so the planes were towed away to a safer spot. Then the US Military thought that the forest where the rails lived could become a potential terrorist enclave from which to attack B52s and wanted it to be burnt down, which would have destroyed the rails. Even Caspar Weinberger heard of the furore, and said the birds should not be harmed. Koko arrived at the base in the front seat of a departmental car. He hopped out and reconnoitred. In the forest, female rails were curious, and males were incensed at another on their territory. Both sexes emerged and were captured.

Now there are about seventy Guam rails and forty king-fishers in captivity, spread between Guam and zoos in America. Neither species is easy to breed. The kingfishers do not like eating zoo food, preferring wild lizards and grasshoppers, but there have been some spectacular breeding successes. Perhaps one day they can be returned to Guam, but not before *Boiga* has been exterminated. There is one fear in everyone's mind. With the increasing air and boat traffic from Guam, it may not be long before the deadly tree snake makes it to islands elsewhere in the Pacific.

Having returned from Moorea to Tahiti, I was woken the following morning to the sound of the Marseillaise. Bastille Day was approaching and French flags were fluttering in the streets of Papeete. Pavements opposite the yacht pontoons and harbour front of the Tahitian capital thronged with excited people as the two-week fête – dancing, feasting, bazaars, canoe races, contests – approached its climax. By eleven 'Le Bar' was filled to the brim, roadside stalls were doing a brisk trade in 'stek frite', noisy motorbikes dodged between colourful lorries converted into buses. The influence of France was everywhere. Shops lined the streets filled with elegant clothes and exotic perfumes that few Tahitians could afford. The fragrances of excellent fresh coffee, *tartes aux pommes*, croissants and *pain au chocolat* filled the air.

France will not give up its Pacific gems easily; on the contrary, unlike all other colonial powers in the Pacific, she has been tightening her grip. Her possessions here include New Caledonia, the Society Islands, the Tuamotus and the Marquesas. The indigenous Kanakas of New Caledonia become more restless each year. Outnumbered by immigrants, they see little hope of a greater say in running their own affairs, short of boycotting democracy and voting with a gun. To the south-east of Tahiti, the small island of Mururoa provides a site for France to test her nuclear weapons. Unlike Britain and America, which have released information on the appalling levels of contamination following tests on Christmas Island, Bikini and other sites in the Marshall Islands, France has kept a veil of secrecy over the radiation pollution that many experts believe must have accompanied their tests in French Polynesia. The USA, Britain and France have exploded about 250 nuclear bombs in the Pacific since the war. The fractured coral core of Mururoa, where the French underground tests take place, is heavily impregnated with plutonium, the deadliest substance known to man. This has a radioactive half-life of 24,000 years, so it seems almost certain that some of it will find its way into the Pacific, if it has not already done so. The island is soon to be abandoned in favour of the neighbouring atoll of Fagataufa.

Kwajalein, on the other side of the Pacific, is the largest atoll in the world. It is used by the US military as a target for intercontinental ballistic missiles fired from Vandenberg Airforce Base in California. The Marshallese who once lived in solitary splendour on the atoll now inhabit the worst slums in the Pacific, on the tiny islet of Ebeye in the giant atoll's ring. The lagoon opposite the settlement, which has a higher concentration of buildings than Hong Kong, has a bacteria count

25,000 times higher than the US Public Health Service demands. Beer cans and 'disposable' nappies float in the bays. Most of the native population now lives on US welfare; before, they needed only to harvest the land and fish, skills now largely abandoned.

Anxious to win hearts and minds, France pours considerable amounts of aid into the islands, most of which enters the pockets of French civil servants on high salaries. In many parts of the Pacific it is easy to see how aid can be as lethal as heroin. The sense of wellbeing young nations feel is soon replaced by a helpless addiction to the cash handouts on which their governments depend and complete subservience to the source of supply – something the suppliers know only too well. Nowhere is this demonstrated better than in the tiny state of Palau in the Western Caroline Islands, which has been subjected to intolerable pressure from the USA to accept giant military bases in preparation for the day when those in the Philippines become politically untenable. Despite repeated referendums among the 15,000 islanders, the proposals were rejected. America withdrew its aid, the Government collapsed and the Prime Minister committed suicide. The islanders live in a state of confusion, but remain adamant.

A gigantic new airstrip in the Marquesas, to which few at present fly, and a geo-stationary communications satellite which has been established in space above the islands – apparently so that the natives can make international telephone calls, something most rarely do – suggest that France has plans for her territories here. Some suggest she is preparing to remove herself from the Society Islands and concentrate her weapons programme in the Marquesas, where the clamour from a mere 7,000 inhabitants will be easier to control. Each year the cry for a 'nuclear-free Pacific' grows louder as fledgling nations, led by Vanuatu and encouraged by New Zealand's powerful anti-nuclear stance, flex their political muscle; each year it becomes harder for France and America to test their weapons in someone else's Eden.

On the streets of Papeete during the fête, none of this was apparent. At nightfall the thrilling sight of Polynesian dancers swaying their hips at breathtaking speed and the sound of native drums pounding a deafening rhythm brought everyone out in smiles. With my blood flowing faster, I retired to a bar. I slid on to a stool next to a tall Tahitian girl in a magnificent white costume of sequins, her hair decorated with flowers. She wore an unusually strong perfume. The night-club seemed to be full of women, somewhat taller than most I had seen, lurid with lipstick and smiling somewhat ravenously at the few drunk French servicemen and tourists who stumbled around in the dark. I turned to the girl next to me and said, 'Were you dancing in the fête?' She turned and looked into my eyes. I noticed she had rather hairy arms.

'No, I wasn't. Hi, my name is Pauline,' she said, and I realized 'she' was a man. The modern Pacific is a mysterious place. Transvestism is a popular tradition in much of Polynesia; a family without a girl may bring a young man up as one to fill the gap. Men dressed as women, known as *rea rea*, serve in many restaurants, but in much of Papeete they seem to have become lurid social butterflies.

One final destination lay before me; one where I hoped all the forces I had witnessed on my journey across the Pacific would come together. The Polynesian colonizing spirit had reached its limit at Easter Island, as had many of the animals and plants from the Asian natural world. Man and nature had existed together here without influence from any other culture. An extraordinary society had grown up which boasted the only ancient written language in Polynesia, still undeciphered, and sculpted mysterious great statues in stone. Here was the greatest Pacific enigma of them all.

Chapter Ten

A WINDOW ON THE WORLD

Easter Island

> As our eyes gradually became trained to dis-
> tinguish art from nature, we perceived the
> white mountain was one single swarm of
> bodies and heads ... half-finished giants lay
> side by side staring up into the firmament, in
> which only the hawks were sailing.
>
> Thor Heyerdahl, *Aku Aku*, 1958

here can be few places in the world which conjure up a greater sense of mystery than Easter Island. For decades the name has been synonymous with a forgotten civilization, a lost people. It is a small island, isolated by thousands of miles of ocean in the Eastern Pacific. All that remains today of the culture which once flourished here is a wasteland of grass and hundreds of enormous human statues staring silently across a boulder-strewn landscape of extinct volcanoes. Easter Island is the most remote inhabited island on earth.

It was born about five million years ago, an outpouring of lava rock from the mid-Pacific rise. It lies 3,700 kilometres from Chile, 1,900 kilometres from Pitcairn. Its triangular shape encompasses a mere 166 square kilometres. That any humans should have found it at all must have been a remarkable stroke of luck or a supreme act of navigation. The first European to stumble across its jagged dark cliffs pounded by huge ocean swells was the Dutch explorer Jacob Roggeveen, who sighted the island on Easter Sunday in 1722 (hence its name). He stayed a couple of days, noted the extraordinary statues, found the islanders given to pilfering, and shot twelve of them before sailing away. The European treatment of the Easter Islanders which followed must rank as some of the worst in human history, and is one of the main reasons why so little is known of the origin of the Rapanui people, as they are called, and their statues.

Following Roggeveen's visit, Cook, the French navigator La Pérouse and others called by and made some fine sketches of the statues as they used to be. Then in 1862 Peruvian Blackbirders arrived and forcibly removed almost half the island's population – then about 3,000 – to work as slaves in guano mines and plantations in South America. Most of the prisoners were from the *Ariki* class, including the King, priests and learned men. Following pressure from the Bishop of Tahiti, the fifteen that survived their ordeal were returned to Easter Island, but they brought a deadly gift: smallpox. Disease ravaged the remaining Rapanui, and by 1877 the total population had fallen to barely 100. In 1883, Chile annexed the island and leased it to a British company whose sheep grazed the hills bare. Today there are fewer than 2,000 Rapanui, farming the island in a simple way and catering for the steady trickle of tourists who have come to gaze at the mysterious statues since Mataveri Airport opened in 1967. The present-day culture is largely borrowed from Tahiti and Chile, and of their once magnificent past the islanders only mutter legends and sing strange chants. To unravel the mystery of Easter Island has become the obsessive goal of anthropologists and archaeologists across the world.

I had crossed the Pacific from Palau in the Caroline Islands, where the huge money stones of Yap had been mined, travelled through Micronesia, parts of New Guinea, the Solomons, Fiji, Samoa and Tahiti in the Central Pacific. Now, in the east of this great ocean, I was flying towards a tiny isolated rock which to the people who had lived there had seemed like '*Te Pito o Te Henua*', the navel of the world. The Rapanui never saw any other land; they believed the moon was the nearest place to them. Up to now all the evidence showed that I had been following in the footsteps of the natural world, which had spread from Australia and New Guinea northwards to Micronesia and eastwards into the islands of Melanesia and Polynesia. Despite winds and currents moving against them, everything pointed to a South-east Asian origin for the majority of animals and plants in the Pacific and also for its people. But Easter Island is 8,500 kilometres from Asia and more isolated than anywhere else in the Pacific. Could the same be true here?

The origin of the statues on Easter Island has been explained by many theories, from the idea that they are remnants of Atlantis to the suggestion that they are the work of beings from other planets. One man threw scientific convention to the wind to prove his point and has stirred emotion ever since: Thor Heyerdahl.

Following his successful journey on a papyrus-reed raft, from South America to the Tuamotus, Heyerdahl carried out extensive excavations at Easter Island in the mid 1950s, and pronounced that the first inhabitants of Easter Island had arrived from South America in a similar way, in about AD 380. A second wave, also from South America, he said, had arrived several hundred years later and destroyed the former colonists by throwing them into a huge burning ditch, evidence for which he discovered near the Poike peninsula, in the east of the island. The Rapanui themselves have a legend that the 'short-ears' overthrew the 'long-ears' at this site. Heyerdahl believed that the victors then entered a great age of statue-building during which they raised the human-looking monoliths or *moai*, weighing 20 tonnes or more, on to ceremonial platforms of dressed stone blocks called *ahu*.

Some of these platforms do bear an uncanny resemblance to the Inca temple walls of Cuzco and Machu Picchu. Most are to be found near to the coast. Strangely, the eyeless statues face inland, their backs to the sea, dominating the people who must have worshipped them in the past. They were gods of a sort, perhaps great chiefs immortalized in stone. Some once had enormous circular topknots or *pukao*, carved in red stone and weighing several tonnes, raised on to their heads.

There is much evidence on the landscape of fighting and war from the late seventeenth century – mainly scattered obsidian arrowheads – which Heyerdahl attributes to an invasion of Polynesian peoples from the Marquesas, who overthrew the South Americans.

Others hold a totally opposing view: that South Americans played no part in the history of Easter Island, and that the original colonizers were Polynesians, perhaps from the Marquesas. The earliest carbon-dated evidence suggests this would have been in around AD 400. According to this theory, there were two main building periods. The *ahu* began as mere extensions of the Polynesian *marae* temples in around AD 800, and had no statues on them at first. Then a second period began in around 1100, when the great statues were built on top of the *ahu* and large rectangular plazas of sea-rounded stones were laid in front of them. The statues were modifications of stone back-rests for chiefs, such as I had seen at Huahine. Statues similar to those on Easter Island, but smaller, are to be found in the Marquesas and other islands of Polynesia. After 1680, no more statues were built. War and chaos broke out, due not to a second invasion, but to over-population.

I was fascinated to know what the island had been like before man had arrived. Was there once thick forest growing over the volcanic slopes where now there is only grass? Apart from a few areas planted with exotic trees by the Chileans, virtually nothing remains of the original vegetation; even the first European explorers commented on the barren nature of the landscape. What kinds of birds had managed to reach the islands, and did they evolve into strange forms as in Hawaii? Had there once been enormous beetles as in Fiji, or huge eels in the lakes at Rano Kao and Rano Raraku as in the Cook Islands? Had the first human colonists consumed the island's wildlife and so modified the landscape that they ended up destroying themselves?

'Does anybody know Edmundo?' I cried amidst the mêlée in the small arrivals hall at the airport. There was a sea of Polynesian faces, many with Chilean looks which reflected their contact with Spanish blood. I had no idea where to go. My contact at the Museum of Archaeology in Tahiti had said, 'Go to the airport and ask for Edmundo. He's an archaeologist who works there but he's not there at present. Someone will pick you up who knows him.'

At the back of the crowd I could see a plump woman smiling and waving frantically. 'Edmundo, he is my friend,' she shouted. I squeezed through the crowd towards her.

'Hello,' she said with a sparkle. 'My name is Siki Raputuki. Edmundo stays at our hacienda; he is married to one of my sisters. You can stay with us.' A wait followed as she looked for a cousin with a truck, then we set off down the new tarmac road to Hanga Roa, the only town on the island.

Hanga Roa almost has the appearance of a shanty: tin houses

line the road interspersed with banana palms and trees. Most of the streets are not paved; they are tracks of red earth and stones over which small horses trot bearing long-legged islanders on wooden saddles covered with colourful rugs. After the opulence of Papeete, I was struck by Hanga Roa's poverty. Mangy dogs and rusting cars shared the sidewalks. Small shops offered tinned foods, and Rapanui women and their daughters sold a sparse selection of vegetables from open shelters. Only one supply ship calls each year, so the prices of consumer goods are high and variety is limited. Jeans and strong boots are often more welcome than currency. Bumping up a small track through the bushes outside the town we turned through a stone wall into a courtyard surrounded by four tumble-down cottages.

'Mama!' Siki cried. 'A friend of Edmundo's is here!' Through a battered wooden door a large Rapanui lady with a beaming smile emerged and gave me a welcoming kiss. Mama had fourteen children, seven of them daughters. At sixty she looked remarkably young, drank vast amounts of black coffee and never stopped talking. With animated expressions and side-splitting laughter during which her ample form threatened to spring from her clothes at any moment, she engaged me in lengthy conversations about her family and the legends of Easter Island, quite undaunted that I could not understand a word.

I met Peco, Siki's husband, a very tall gangling fellow with a mass of frizzy hair and a face covered in freckles, and Arturo, married to beautiful Ortesia, and so it went on. Finally Mama announced that my arrival was perfectly timed because it was the day for the all-night disco, and I could accompany her and the family to it. With that she announced that we should walk down to the coast to Ahu Tahai, the nearest of the statue monuments.

There is something remarkably powerful about the Easter Islanders' images of man. Though I had seen them many times in photographs and on television, I felt a shiver of unease when I first gazed on them face to face. The five statues, each one four times my height and so wide that I could not stretch my arms across its belly, stared across their courtyard of smooth, round stones towards the grass-covered hills in the distance. It was as if they ignored the might of the ocean behind them and were concentrating on some great event taking place inland. Their heads had fine lines to the jaw and brow, and deep sunken eye sockets which had once been filled with white coral eyes. At either side of their heads, ears with long stretched lobes hung down, flanking wide, generous noses beneath which thin lips were pursed in permanent silence. From their shoulders long arms were carved into their sides, with hands stretching on long fingers towards their navels. Each statue was almost identical.

Around the island archaeologists have so far discovered 745 *moai* in nine basic forms, all very similar to each other. The majority lie toppled from the *ahu* where they used to stand, though a number have now been restored. Many statues lie broken where they fell in the process of being transported across the island from Rano Raraku, the huge volcanic crater on the eastern tip of the island where they were made. I decided to journey there on horseback, crossing the northern coast; on the way I would see *ahu* and other monuments of Easter Island's turbulent history inaccessible by road and perhaps discover some of the ancient clues to its past.

That night at the disco I was a star. The only outsider in the place, I kept being scooped from my chair by sometimes less than ravishing Rapanui girls to gyrate vigorously around the floor to Polynesian pop. There was no let-up, save when a girl's ample mother cut in, clasped me to her capacious bosom and wheeled me round.

Nursing aching heads, Peco, Arturo and I clopped out of the courtyard on three small horses, followed by Hueso, their large sandy-coloured dog. Chilean horses trot with a rapidity which in a normal saddle would be bearable, but in one made of wood soon becomes agony. Peco clung to a huge bundle tied up with plastic which contained our bedding, Arturo managed pots, pans and food, and I carried cameras and clothes in my rucksack. Under the brightening sun we trotted down small red tracks between walls of boulders before emerging into the wide open grasslands to the north of Hanga Roa. Of native wildlife there was almost nothing to be seen. A few small grey birds introduced from Chile flitted among the trackside bushes. 'Manu toke-toke.' Arturo pointed to the sky. A handsome buzzard wheeled over our heads on majestic wings. There were a great many of these on Easter Island; another Chilean introduction. They had a habit of alighting on stone boulders among the grasses, bowing and throwing their heads back in a wheezing screech as though beside themselves with mirth. No native land birds nor their fossils have ever been discovered on Easter Island. Perhaps it was just too isolated for even birds to reach, or perhaps the first islanders so efficiently exterminated them that not a trace of a pigeon or a parakeet is left. The same is true of most other animals and plants found on the island today. Few show any connection with the Marquesas or Tahiti. Some certainly came from these places but others are weeds and appear to have colonized the island since European times. Some hard *toromiro* wood, much favoured by the islanders for carving, still existed on hills when Roggeveen surveyed the island, but the trees were finally destroyed by sheep, who removed their bark, in the nineteenth century. Now only a few stands exist on Rano Kao crater. It seems that all the original vegetation of the island has been replaced by colonists.

Peco stopped by a bush and leaned from his horse to pluck several ripe guavas and we rode on. Arturo cantered in from over a hill where he had found some lemons. Easter Island is too cold for corals to grow around its shores or for coconut palms to clothe its rocky slopes. The drying winds that sweep across the hills made it almost impossible for the castaways that landed here to cultivate crops. From time to time we passed small circular stone walls surrounding pits in which banana palms and taro grew sheltered from the wind. Arturo stopped at one to fill our water bottles from a pool in a cave leading off from the pit. The island is riddled with caves. Many of them contain the remains of human bones and artefacts, indicating that once they were lived in. The caves are old lava tubes, a legacy of the island's volcanic past. Pulling up my horse, I crawled into one through a break in the roof. Inside was a small room faced with dressed stone. Sunlight filled it through the hole above, lighting the green-baize moss which coated the damp walls. When Cook called at Easter Island he thought the population must be predominantly male, not realizing that most of the women were hiding underground.

We paused for lunch at Ahu Akivi, a magnificent restored

ahu of seven *moai* set in an impressive amphitheatre of rolling hills. Moving on we came across long oval-shaped foundations drilled with holes about the size of ducks' eggs. Into these holes, hoops of timber had been placed overlain with thatch: the original surface homes of the Easter Islanders. Each of the oval shelters had one small door in the side and could house perhaps a hundred people huddled together. Some of these foundation stones had been incorporated into the secret cave house, so whoever built it did so after the oval shelters had fallen into disuse. Something had driven the Easter Islanders underground. Was it the fear of a volcanic eruption or an upheaval of a different kind? In some places on the island, small rounded houses of stone blocks, like miniature fortresses with small entrance doorways, were built close to settlements. They were too small for humans. Their use for years remained a puzzle until bones inside them were analysed and found to belong to chickens. They were fortified chicken-houses, and dated from the eighteenth century. Chickens had clearly become immensely valuable. There were no predators on the island, so someone must have been interested in stealing them, necessitating heavy protective measures – but why?

My face was burning from a day in the sun. Thwacking the horse with a twig, Peco grinned and whooped as he and Arturo raced across the hills of Easter Island leaving the setting sun behind them. We had turned the northern tip of the island, and now approached the steep hills of Hangao'tea where we were to spend the night. In a small rocky bay liquid lava had once spilled over a cliff into the sea. Now only a tube-like cave remained, providing a perfect place to shelter. Hueso slumped in the dried grass that lined the cave floor, and coffee was soon brewing on a fire. In a eucalyptus grove near the cliffs behind us, forty or fifty buzzards gathered to roost, bowing and wheezing as darkness fell. After a supper of tinned meat and soup, I collapsed into my sleeping bag and slept the sleep of the dead.

'Wake up.' Peco was shaking my shoulder. 'You want to come fishing?' It was still dark; only a few hours had passed. We crept into the night and stumbled across the rocks near breaking waves which signified the shore. There was no moon, and our torches flickered across the foam. Peco caught some crabs in pools among the rocks and crushed them with stones, tossing their remains into sea, then threw in a line baited with fish. We waited. Hueso whimpered behind us. The sea roared in the darkness, splashing phosphorescence on the rocks. A shooting star sprang across the sky.

'*Huevos!*' Arturo tapped my shoulder and pointed to the stars. I had no idea what he meant. '*Hu-e-vo-s.*' He pronounced it emphatically.

I got it the second time. 'Ah! UFOs.'

'It's normal here,' said Arturo. Ashes from his cigarette flew on the wind like meteorites.

We returned after an hour with a grotesque *kopuku* and a large *nanue*, a typical night-shift fish, red with large eyes. The first colonists here would have found these shores barren in comparison with the rich coral reefs of their tropical homes. Archaeologists have unearthed many fish-hooks of bone and stone on Easter Island, indicating a move from offshore to inshore fishing as the colonists adapted their skills to the coasts. The hooks are similar to those used by Polynesians on islands further west; an important clue when one comes to guess at whose hands were making them.

The following morning we set out early, hoping to reach Rano Raraku by nightfall. The horses' hooves clattered through the lava stones, following a worn coastal path beside steep cliffs. A whale spouted among the swells, the sun catching the spray, before its tail flukes disappeared under the blue surface. Near to the cliff edge Peco jumped from his horse and pointed to an area of flat rock. 'Petroglyph,' he explained. Just discernible in the stone were markings and lines about the thickness of my finger, but it was difficult to say what they depicted. Peco pulled out a bottle of baby powder, mixed some of the contents with water in a small bowl and began tracing a white outline on the rock with a small paint brush. He was used to working with archaeologists on the island. Gradually the image appeared of a peculiar bird-like head with enormous eyes and a large hooked beak attached to the shoulders of a human form which was kneeling and holding out a circular object in its hands. There are a number of these bird-men on rocks along the path. One pattern, almost a metre long, was stranger than all the rest. It was only when Peco had finished with the powder paste that a form appeared with legs and arms and a large round head, dressed in a kind of bulky suit with straps. I gazed at it from all angles, not wanting to accept what it clearly looked like.

'Astronaut!' Peco laughed, showing off his stained teeth, and sat back on his haunches looking up at me, his hands on his knees. Perhaps it was just the way he had drawn it. Rain would soon wash away the white paste from the rock, leaving the pattern for others to interpret.

In a few hours we reached Anakena Bay. Here legend has it that the ancestral King of the Easter Islanders, Hotu Matua, arrived by canoe. The genealogies of Easter Island kings place this event in around the twelfth century. Anakena is certainly the most beautiful bay on the island, the more so for a magnificent *ahu* supporting a number of statues restored to their former glory with the huge red *pukao* topknots on their heads. In 1956 Thor Heyerdahl demonstrated how a statue could be raised into an upright position here using ropes and beams and a ramp of stones to support its weight. But how the heavy *pukao* was placed on the statue's head still remains a mystery.

We rode on from Anakena past the bay where La Pérouse first landed more than two and a half centuries ago and turned south beside the extraordinary Poike Trench, visible as a slight depression now, separating the west of the island from the mountains of the peninsula. Modern archaeologists regard the ditch as a modification of the natural landscape used for growing crops such as taro, rather than as the Easter Island equivalent of the Maginot line, as Heyerdahl believed. At Tongariki on the south coast the largest *ahu* of all now lies in ruins. Here twenty-tonne *moai* lie a hundred metres inland from where they once stood on their stone platform beside the beach, scattered like so much driftwood by a powerful tidal wave which struck the island in 1962.

A short distance to the west of Tongariki, the huge volcanic crater of Rano Raraku rose from the dry grass plain into the afternoon sky. We rode over its southern flanks and I began to see the outlines of numerous stone statues, buried to their shoulders, sticking out of the sea of grass like giants struggling to be free. In the great statue-building period, hundreds of islanders would have been employed inside the crater, using stone scrapers to carve the figures from tufa, a form of compressed volcanic ash. Some had to be abandoned half-way through when fissures in the rock spoilt their forms or when they were too hard to work. Sometimes they simply cracked before they could be removed. A statue would be gradually hollowed from the rock in a lying position, its spine alone holding it in place. When the time came to move it, small stones would be placed under its back, the spinal attachment chipped away, and the statue slid down the mountainside, carefully controlled with ropes. It would then be slipped upright into a pit to await transport.

Nobody knows how the statues were moved to their final places of worship, which included a site at the far end of Easter Island, over 20 kilometres away. The remains of a road exist, but no clue as to how they were transported along it. Some weigh more than 60 tonnes. Easter Islanders themselves believe they were transported to their *ahu* by magic. Professor William Mulloy, one of the fathers of Easter Island archaeology, proposed that they could have been suspended belly-down on ropes and swung through an inverted V-shaped timber cradle a few metres at a time. Heyerdahl showed that 150 men could pull a statue on its back along timber rollers – though there is no suitable timber left on Easter Island today. Others propose that they were wobbled upright from side to side by ropes attached to their heads – which would fit legends that the *moai* walked to their *ahu*. Whatever the method, it almost certainly entailed huge amounts of manpower over a long time, which suggests a great devotion to the statues and a powerful hierarchy capable of marshalling the necessary labour.

Inside the crater of Rano Raraku I could see almost 400 statues, in various stages of completion, waiting to be transported to their *ahu*. It may have become impossible to transport all the statues that were being made – owing to a lack of wood on which to roll them, if that is indeed how they were transported. There is no history book to which we can turn to tell us whether trees once clothed the island's barren slopes and what happened to them. Fortunately, there is another medium to provide an accurate account of Easter Island's botanical history, and one which goes back thousands of years. It is to be found in the bottom of Rano Raraku's lake.

Pollen makes an excellent record of the past. That blown on to the water of the crater lake inside Rano Raraku and trapped in layers of sediment records the plants which once grew on the island as accurately as the pages of a book. On the Scientific Exploration Society's expedition to Easter Island in 1986, Dr John Flenley of the University of Hull took a mud core from the lake and attempted to identify the pollen grains found in it. The results were startling. In the deepest parts of the core, grains of forest species no longer here were present in large numbers. Evidently the island *was* once covered in trees. Other absent species such as Compositae, similar to sunflowers, and the Polynesian rope tree were also originally present. Coinciding with the time of man's arrival, new species of pollen belonging to crops such as taro and bananas and to grass began to appear. Gradually the forest pollen declined, and grass increased, until the forest pollen had all but vanished and crop pollen declined, leaving grass to reign supreme. Charcoal in the soil shows that the Easter Islanders burned the landscape on a grand scale, perhaps to create farmland; a scheme which served them well in the fertile tropics from where they had come but here proved disastrous. The decline in the crop pollen appears to coincide with the period of greatest conflict on Easter Island.

One find in the cores was particularly intriguing. Large numbers of grains appeared to belong to a palm; but not *Cocos nucifera*, the coconut palm, which is the only one existing on Easter Island today. There are approximately 2,700 different kinds of palm. To identify this one from pollen alone would have been impossible. A remarkable find in a lava tube of some ancient palm nuts narrowed the field: these were very similar to those of the Chilean wine palm *Jubea chiliensis*. It seems that Easter Island evolved its own species of palm that once may have covered large parts of the island, but today not one is left. The Chilean wine palm is enormous, reaching the height of a four-storey building; its trunk is smooth and almost perfectly round. The Easter Islanders could not have asked for a tree better fitted to roll statues on. Their own palm was almost certainly similar, and it appears they cut every last one.

The lake has another secret to offer. It is set at the bottom of a great amphitheatre of pale-brown rocks, many hundreds of metres across. Green spiky reeds grow around the water's edge; these have their nearest relatives in South America, where mainlanders use them to thatch houses and even make rafts. Thor Heyerdahl believed South Americans had planted them on Easter Island. The fact that the sweet potato, a native of the Andes, is also found here was another important part of his argument that the original inhabitants of Easter Island were native Americans, but the fact that many of the ancient crops that occur on the island are similar to those of Polynesians argues against this. The ancient language of the Rapanui is an Austronesian tongue similar in origin to that of Polynesia and most closely allied to Marquesan. Linguists believe that about 19 per cent of the spoken vocabulary of a split people diverges from its origin every thousand years. That would place the split between Rapanui and Marquesan at about AD 400, roughly the time man is believed to have first reached Easter Island. Ancient Easter Island skulls provide another clue: their noses and jaw bones bear a much closer resemblance to skulls of Polynesians than to those of South American Indians. Taking the evidence together, it seems almost certain that the original colonizers of Easter Island were Polynesians, probably from the Marquesas.

High on the edge of the volcano, I looked down the grassy sweep of the bowl where the statues lay and across the circular lake, right over the island past many smaller volcanic cones dotted with black rocks towards Hanga Roa. Rising to the left of Hanga Roa was the only other massive volcanic crater on the island, Rano Kao, tomorrow's destination.

As I walked out of the crater towards the camp on its southern flank I came suddenly upon a gigantic statue, its torso more than 20 metres long and weighing more than 200 tonnes. It lay on its back, half-finished, and surely could never have been moved. I walked up its belly to its eyeless head: it was a fine monument, a fit image, for a people who could not see. What foolish king had commanded it to be built? Was this the final straw that caused an uprising among the serfs required to hew these massive forms long after the means to move them had been destroyed? This was a culture obsessed with itself: even the fact that the *moai* stand on their *ahu* with their backs to the sea, looking in, seems to sum up the nature of a people unaware of the world outside. In the gathering darkness even the buzzards were silent now.

The great crash came around 1650, when the island was thrown into war. While this may have been an uprising of workers against their masters, as ancient legends hint, it was almost certainly fuelled by the fact that the landscape was now almost bankrupt of natural resources, for which competition became keen. Why else would chickens be housed in fortresses? Conflict between the haves and the have-nots became inevitable, and since the number of have-nots generally exceeds that of the haves it was only a matter of time before the former triumphed. What followed was the dark age of Easter Island. A great industry developed in the manufacture of weapons, principally the obsidian arrowheads which now lie scattered all over the island. Human beings were no longer able to live in peace above ground, so they crawled beneath it to shelter their wives and children in caves.

It took most of the morning to ride the length of the southern coast; from here we began to climb the slopes of Rano Kao through plantations of eucalyptus and pines. The crater loomed unexpectedly large. I was able to look 200 metres down the near vertical cliffs to the mysterious lake at the bottom. The interior is precarious to reach; here alone some of Easter Island's original plants have survived. The opposite side of the rim dipped where the sea had almost breached it. We cantered through the mist, descending a little to the sacred village of Orongo. Here, overlooking sheer cliffs to the sea, was a collection of houses made from flat stones placed one upon the other. Small doorways led into them, and inside were bare mud floors. At a point on the cliffs opposite two small islets offshore, the rocks were covered in numerous petroglyphs like those I had seen on the northern side of the island. This was the centre for the cult of the Birdmen, which developed out of the chaos of the old order on Easter Island.

At the time of the first European contacts with Easter Island the Birdman cult was still being practised at Orongo, so history has passed on a few tantalizing insights into its meaning, even though the Peruvian Blackbirders all but wiped it out. There are some 6,000 petroglyphs on Easter Island. Many depict the frightening Make-make god, turtles, lizards or fish, but many conform to that of the Birdman. I found most in the village of

Orongo where a great annual ceremony used to be held: priests and chiefs would gather here to choose the Birdman for the year. The heads of the crouching figures in rock that I had seen looked like the heads of frigate birds, and the round objects in their hands were eggs. Each year the priests waited for the time when sea birds would return to breed on the small islets of Motu Nui and Motu Iti opposite Orongo. A great gathering would meet at the village and the best swimmers, or *hopu*, were chosen to climb down the sheer cliffs and swim to the islets. There they would wait, perhaps for many days, for the first eggs to be laid. The first *hopu* to find one wrapped it around his head in a piece of red *tapa* cloth and made the dangerous swim back to the shore, through the breakers and up the cliffs, to present it triumphantly to his Chief. This Chief was then ceremoniously appointed *Tangata Manu* or Birdman, had his head, eyebrows and eyelashes shaved, his face painted in red and black streaks, and spent the rest of the year in isolation in a hut at Rano Raraku, sacred until the next ceremony.

The *hopu* did not go unrewarded: he was given a virgin as his prize. I noticed footholds carved into two large boulders on either side of a narrow passage at Orongo. To ensure that the young girls who presented themselves were in fact virgins, each in turn would step into these, enabling a priest to walk between her legs and make an inspection. Lest his knowledge of female anatomy should fail him, small diagrams on the boulders nearby demonstrate the difference between a virgin and a non-virgin in impressive detail.

Some aspects of this story are confusing. Frigate birds are rarely if ever seen around Easter Island's shores. Why should a frigate have been chosen to represent the Birdman? It may be that the birds which visited the islets offshore were in fact sooty terns, and that it was their eggs the *hopu* sought. On the other hand frigate birds have a strong homing instinct. It may be that they were carried on Polynesian migration voyages to aid the search for land. They could even have been used to carry messages back to home islands after long sea voyages. For whatever reason, frigate birds appear to hold a special place in Polynesian society, and this alone may be why they were chosen. The Easter Islanders had destroyed their natural world:

virtually all the original trees and shrubs which clothed the hills had been erased. Any wildlife, in the form of pigeons or other birds suitable for food, would have gone with them, though we shall probably never know whether such birds existed here. Without timber, the Rapanui were unable to build boats to undertake long ocean voyages in the tradition of their forefathers; they were imprisoned on a devastated island of their own making.

Sea birds almost certainly used to nest on the cliffs of Easter Island and would have been a valuable source of food. The islanders banished these too: only on the tiny islets off the south-western tip did they survive. It seems to me that these birds represented to the islanders everything they had lost. Unlike the Rapanui, the birds were free to move across the ocean. Their return each year renewed the islanders' faith in nature and the rewards it could bring. The Birdman's egg represented something far more valuable than anything the islanders could fashion with their hands, and became the cornerstone of their spiritual beliefs. I wanted to visit the small islets or *motu* where the birds had bred to see if they returned there still. From the cliffs of Orongo, a few birds appeared to be flying about the rocky crags, but it was impossible to see what they were. I was leaving the following afternoon; there would just be time to visit them.

At nine o'clock the following morning, sore from the previous three days' riding, I arrived with Siki and Peco at a small harbour in Hangaroa. 'A cousin' agreed to take us out in his sturdy, clinker-built, open rowing boat. It was about five metres long and painted bright orange inside, with a suspect-looking outboard on the transom. As we set out to sea a light drizzle began to fall. At the harbour entrance, large angry swells funnelled between the rocks, making us pitch in an alarming way. They were bigger than anything I had experienced in the Solomons before, and soon we were drenched with waves. We entered a calmer sea and moved south alongside the imposing cliffs, which were layered with successive falls of brown ash like a cake. Passing the shore I imagined a small figure clambering up the huge black cliffs, a single egg bound to his forehead. Great humps of water bounced back from the shore where the waves met the cliffs. To land there seemed madness.

Motu Kao Kao rose like a damaged and stained dog's tooth from the waves around its base. It was the smallest of the three islets, and its summit was coated with an icing of bird droppings. We skirted Motu Nui, the largest islet, about fifty metres high and several hundred across. The boat bucked in the confused sea as we motored round to the sheltered side. As we approached the shore the swells rose and fell leaving the rocks pouring with foam, like blackened teeth with saliva gushing around them. Timing was all important. As the boat nosed in, I leapt ashore and rejoiced at feeling my shoes grip on the rocks. The boat came around again so that Peco could land, my cameras strapped in a bag to his back. A huge, unexpected swell caught the boat and began to surf it towards the shore; the outboard was slammed into reverse, but too late. With a crash the bow cannoned into the rocks and stuck fast. I watched with horror as the hull turned upwards and green water poured over the side, pitching Peco, Siki and the driver into the sea. Then the boat came free, sliding violently down the rocks as the water drained away, and slipped beneath the waves.

Another swell was gathering as Siki floundered near the rocks. I sprang down to the sea and grabbed her outstretched hand, pulling her towards me as the wave roared in, clinging to the rocks as the foam swirled around threatening to tear us free; then we scrambled clear. Fortunately the boat had not sunk but wallowed with a fraction of freeboard to spare. Peco clung to its side, supporting it as his cousin pulled himself carefully aboard. Unbelievably, the outboard started, and before a third wave came, he chugged the boat out to safety. Siki and I collapsed in nervous laughter at our narrow escape, then I quickly explored the island. I found some petrels, one broken bird's egg and a few pieces of obsidian. The buzzards laughed at me from the rocks; they would reign here now, until the sea birds came again to breed. We boarded the boat, and left.

The shore was welcome after the grasping waves of the harbour entrance. There was time enough for one of Mama's vast lunches. 'Eat! Eat! The plane will wait,' she said, spooning vast pork chops on to my plate. At the airport the family weighed me down with shell necklaces and sad goodbyes before the ageing Lan Chile 707 lifted off. I strained my eyes to see a last glimpse of Ahu Tahai, but it was hidden in mists.

Across the Pacific each island has been a small experiment in nature, of which man is just a part. Only on Easter Island was he as isolated as a bacterium in a bottle of broth. His progress there almost exactly followed the line the bacterium would take; building to a high population beyond means of support, using up all supplies, then crashing into a cesspool of his own making. Here is an island in the vastness of the Eastern Pacific, isolated as our world is in space. Evolution provided it with a collection of original animal and plant colonists which harmoniously developed the island's natural resources. Then the island was seeded by a small population of human beings. Initially they lived frugally, adapting the skills they had learned on their island home to the new conditions they found. Rapidly they evolved a culture of their own. The Easter Islanders burnt their forests to the ground as the population grew at its height to around 12,000 people. They became obsessed with a material culture fashioned by their own hands, perhaps as a way of employing surplus labour. They, as we are doing on a world-wide scale, ignored a fundamental law: that a varied natural system is inherently more stable than a monoculture, so that when one part fails another may support both. They ignored the fact that their lives depended on the resources provided by their natural world. Without timber to move the stone idols, the statue industry was bound to fail. Once the forest was cut and soil laid bare to the drying wind, crops became harder to grow. Protective walls had to be built around restricted water supplies. They unthinkingly cleared what remained of the forested landscape to feed themselves. The Easter Islanders hurried towards their doom. It is perhaps no mystery that today they revere objects carved in wood: strange human statuettes, with protruding vertebrae and ribs, as thin as starved skeletons.

Yet on Easter Island some people did survive the holocaust and evolved a new purpose. The grandest expression of Polynesian material culture had destroyed itself through a disregard for the simple laws of nature. In a landscape starved of variety, a new hope evolved based entirely around one symbolic natural event: the arrival of birds and the renewal of life. How curious it was to find, at the very end of my journey, a link to its beginning. Perhaps Nauru, where the seabirds no longer come, is a modern-day Easter Island.

The Pacific Island cultures are strong in many ways, but they are now being exposed to unprecedented change. Powerful nations have woken to the fact that the Pacific covers half the globe, and find its peoples malleable, even expendable. The oceans offer great mineral wealth and a marine harvest of unprecedented proportions, but it is not the islanders who will grow rich on the spoils of their ocean world. Soviet and American submarines play cat and mouse around mountain ranges beneath the sea, and Japan plans to dump nuclear waste in the Mariana Trench. Vanuatu's Ambassador-at-large, Barak Sope, voiced this popular sentiment:

> In the past the colonialists wanted our labour, so they kidnapped us. Then they wanted our land, so they stole it for their plantations . . . The Trident submarine may be a far cry from the old blackbirding vessel, but to us they are all ships from the same old fleet.

Agreements such as the 1985 Treaty of Rarotonga serve a useful purpose in attempting to unite this great realm of separate nations, for it is only in this way that they will resist powers greater than themselves. While they remain vulnerable to the aid weapon, however, they must be prepared to swallow a bitter pill. Tourism is likely to provide an economic cure but it too may prove a dangerous panacea.

All is not yet lost. Melanesian and Polynesian culture has resisted invasion in the past and is still vibrant in a way that encourages you to believe that dreams can come true. The worst fear of the Pacific Islanders has been realized in Hawaii, where a once proud culture has vanished beneath a sea of Western-led consumerism. Careful management of tourism will bring great benefits to the islands, but the lives of their people will not remain the same. The fact that there are so many endangered species in the Pacific is a draw to tourists, as are the rugged wildernesses which sustain them. Both wilderness and wildlife are in urgent need of protection if they are not to become part of the Pacific's ancient history, and if the islands are not to lose one of the greatest renewable assets they have to offer an inquisitive world.

The greatest threat to these magnificent island landscapes and the fascinating creatures they contain – the destruction and burning of the forests on the high islands – also directly contributes to the pall of smoke and carbon dioxide from other parts of the world that is already causing the temperature of the earth to rise. This atmospheric warming of between 1° and 2°C will inevitably cause ice at the poles to melt and the sea level to rise, as it did at the end of the last ice age. If current calculations are correct, the sea is beginning to rise at a rate faster than corals can keep pace with. Reefs requiring shallow water may become less productive, or even die. Of still greater concern is the fact that the ocean may rise by as much as 1.5 metres in 70 years. Many atolls are barely more than a metre or two high. A storm surge drawn into the vortex of a cyclone by low pressure is capable of raising the sea level by a further two metres or more. If an atoll is struck, its fresh-water supplies will be contaminated, taro pits will flood, and storm waves will sweep palms from the land. In less than a hundred years atoll states such as Kiribati, Tokelau and Tuvalu could vanish beneath the sea. These are not idle predictions. The havoc we are wreaking on our landscape is already showing itself in the erratic changes of regional and even global climate which are increasingly apparent today.

Perhaps we should see Easter Island as a window on our world. Would that eyes as wise as Darwin's could look through it to our future; would that he could warn us of what he saw.

A GUIDE TO NATIONAL PARKS AND PROTECTED AREAS

Throughout the 29,000,000 square kilometres of the South Pacific there are more than 2,000 different ecosystems, spread through just 500,000 square kilometres of land from giants such as New Guinea to tiny volcanoes and atolls. There are many places you can visit where it is possible to roam through enchanting forests, or dive amongst spectacular corals and experience islands where perhaps 80 per cent of the species you will see are unknown anywhere else in the world. As tourism grows, so will demand to see these creatures and the habitats they live in, but as each year they dwindle, action is needed now to ensure that they continue to survive and draw the curious to the Pacific.

Excluding Hawaii, only 50 islands have any protected areas whatever. Half of these are remote uninhabited islands, which represent less than a quarter of those of real conservation interest. There are only 19 National Parks of more than 10 square kilometres. Though many areas are in theory protected by law, the difficulties of policing boundaries and monitoring activities within reserves mean that few in fact are.

Agencies such as the World Wide Fund For Nature (WWFN) and the International Union for the Conservation of Nature and Natural Resources (IUCN), have found it cumbersome to deal with so many different Governments for a relatively small land area, and so have a poor record of support in the South Pacific. The governments of the region perceive more important priorities than their natural heritage on which to spend their limited resources.

To create reserves without allowing for traditional or 'custom' usage of the land concerned alienates the local constituency. No conservation scheme ever succeeds without its support. New initiatives are perhaps needed which break free of the western concept of pristine parks which exclude humans unless they are tourists; to be more in keeping with Pacific people's needs. Australia and New Zealand must play an increasingly influential role in supporting conservation measures in the region, as must France. Mindful of the desperate situation in Hawaii, the US Nature Conservancy has opened up several reserves there, and has recently opened an office there too. Japan continues to export tourists to Micronesia and beyond, and imports vast amounts of natural resources from the Pacific, but to date has done almost nothing to support conservation measures there. The South Pacific Regional Environment Programme (SPREP), based in Noumea, capital of New Caledonia, was the only place I visited which offered an overview of the whole region and is a magnificent source of information. It is funded by the United Nations and other aid bodies, including WWFN.

The major international environmental and conservation agencies must invest more in the protection of these islands. Theirs is the challenge of creating the nucleus of a protected area system which most Pacific nations cannot afford. Governments have the responsibility of using their resources wisely, developing forests without destroying watersheds, exploiting fisheries without extinguishing them, encouraging tourism without letting it engulf island life. This sustainable development of the Pacific's resources for its people will encourage generations of its children to remain there and keep *Vanua*, the Pacific Way, alive.

The Third South Pacific National Parks and Reserves Congress in 1985 agreed on three objectives:

- to establish at least one protected area in each country
- to double the existing number of protected areas
- to double the number of ecosystem types receiving some protection.

Since then only one new reserve has been created, on Tonga. At the time of writing these are the most notable protected areas from west to east across the Pacific:

Northern Mariana Islands (Maug and Sariguan Islands only): Notable for the rare *Pteropus marianus* flying fox, seabirds and the Micronesian incubator bird *Megapodius laperouse*.

Guam: None. Protection needed for the few natural areas left. Remaining endemic birds severely threatened by introduced brown tree snake *Boiga irregularis*.

Palau: The Ngerukewid or Seventy Islands Reserve only. A magnificent collection of mushroom-shaped islands subject to ancient taboos. Dugong *Dugong dugon*, megapodes and Palau scops owl

Pyrrhoglaux podarginus recorded. Some of the world's richest coral reefs are here; none are protected.

Federated States of Micronesia: No protected areas at all. The high islands of Pohnpei, Kosrae, Yap and Tol are of greatest interest and suffer greatest risk.

Marshall Islands: Only Pokak and Bikar atolls have any protected status. Large seabird colonies.

Papua New Guinea: A vast and diverse island with 6 Parks, 3 reserves and 11 Wildlife Management Areas. More are needed on New Britain, Bougainville, Fergusson, Manus and New Ireland as well as on small islands with a high degree of endemism. The largest National Parks are McAdam and Varirata. Birds of paradise are to be seen at Baiyer River Sanctuary, as well as opossums and kangaroos. Tonda Wildlife Management area in Western Province has abundant wallabies, rusa deer, bandicoots, giant monitor lizards, numerous birds, crocodiles and fish. Bensbach Lodge provides comfortable accommodation.

Nauru: None. Most natural vegetation destroyed by mining guano.

Solomon Islands: Kolombangara Forest Reserve and Queen Elizabeth National Park on Guadalcanal only. Both are heavily degraded. As logging of the native forests proceeds apace, a National Conservation Plan is urgently needed to protect watersheds and ensure survival of the islands' spectacular wildlife.

Vanuatu: President Coolidge Reserve and Million Dollar Point Reserve – created in 1983 to protect a magnificent wartime wreck. Excellent diving. Duck Lake on Efate has been proposed for the protection of waterbirds and endemic fruit doves. Reef Island has been proposed as a seabird- and turtle-breeding sanctuary.

New Caledonia: An extensive series of 14 Parks and reserves including the most notable marine reserves in Oceania. The region is remarkable for its collections of strange relict plants and reptiles. Parc Territoriale de la Rivière Bleue, home of one of the world's most endangered birds, the kagu *Rhynochetos jubatus*, is fascinating. The bird – New Caledonia's national bird – resembles a heron, but is a species in a family all of its own, and in the morning calls like a dog with a barking howl. Perhaps a few hundred exist in the wild.

Australian and New Zealand Territories in Oceania: Lord Howe, Norfolk and the Kermadec Islands are already protected. All are vital refuges for endemic birds and other species in danger of extinction.

Tuvalu: None.

Wallis and Futuna: None.

Tokelau: None.

Kiribati: Birnie, Kiritimati (Christmas Island), Mckean, Malden, Phoenix, Starbuck, and Vostock atolls in the Phoenix and Line Islands. Some of the Pacific's largest colonies of seabirds are here, including petrels, boobies, shearwaters, frigate birds and terns. There are also important marine turtle sanctuaries.

Fiji: As a result of the coup the Government's excellent parks and reserves plan is slow in coming to fruition. There are 12 reserves including the Crested Iguana Sanctuary on Yaduataba. The *dakua* tree, *Agathis vitiensis*, from which massive Fijian war canoes were once fashioned, no longer exists in Viti Levu, and the last few reserves on Vanua Levu are already being logged for sale to New Zealand. In ten years these great trees will be a memory. Mongoose-free islands such as Taveuni, Ovalau, Kadavu and Gau, which have a variety of spectacular endemic birds, need more adequate protection. The land-scape of some islands looks Scandinavian; this is a result of planting by the Fiji Pine Commission, which hopes to provide all Fiji's softwood needs in perpetuity and remove pressure from the remaining native forests.

Tonga: Four marine reserves and two protected lagoons. The recently formed Vaomapa Nature Reserve is an improvement. The red musk parrot *Prosopeia tabuensis*, imported from Fiji, still survives on Eua. Visit Kolovai village to see the huge bat colony. The incubator bird *Megapodius pritchardii* still lays its eggs in the volcanic soils of Niuafo'ou's crater. The species is protected by law but there is no enforcement. The blue-crowned lory *Vini australis* has vanished from most of the Tongan Islands this century due to introduced black rats. The precise distribution of the banded iguana *Brachylophus fasciatus* remains unknown. The Tongan ground skink *Eugongylus microlepis* has not been collected since 1839.

Niue: One of the few places to have maintained traditional *taboo* areas which act as protected zones.

Western Samoa: 1 National Park, 3 reserves. O Le Pupu-Pu'e on Upolu's northern coast is the finest National Park in the South Pacific. Huge congregations of enormous flying foxes emerge here at night to flap over the forest in search of fruit. 31 of the 37 land bird species found in Western Samoa are unique to these islands. The Tusitala Reserve outside Apia provides a wonderful introduction to Samoan forest birds near the grounds of Robert Louis Stevenson's home at Mt Vaea. A major new reserve is planned on the flanks of Mt Silisili on Savai'i. It is badly needed to protect the tooth-billed pigeon and other indigenous species. Large-scale felling and the planting of exotic teak and mahogany, which native animals cannot use, threatens these forests. In late October or early November Samoans wait to gather the caviar of the Pacific; the writhing bodies of the *palolo* worm *Eunice viridis* rise from the reefs on a specific night to spawn. Palolo Deep Marine Reserve on Upolu is named after them.

American Samoa: Rose atoll and Fagatele Bay on Tutuila only. US Fish and Wildlife Service reports state that the most valuable forest remains on the Manu'a islands. At Rose atoll 312,000 seabirds reside: 97 per cent of American Samoa's population.

Cook Islands: Suwarrow atoll only. A reserve is urgently needed on Rarotonga, the main island, for the delightful Cook Island flycatcher, thought to be extinct but re-discovered in 1973.

French Polynesia: No clear overall environmental policy has existed here for many years though there has been some ad hoc protection. It is an outrage that there are no official terrestrial reserves in the Society Islands. Moorea, Rapa and Nuku Hiva are all greatly at risk. The Society Islands pigeon *Ducula aurorae* is making its last stand in the Papenoo Valley on Tahiti. The Marquesas, with their unique flora and fauna, have just four – inadequate – reserves, on Ei'ao, Hatutu, Ilot de Sable and Mohotani. Fenuaura and Taiaro atolls are also protected.

Pitcairn Islands: None. Henderson, the least altered island in the Pacific, contains examples of species similar to those long since extinguished from other islands, yet it still does not have a protected status. An American millionaire recently proposed carving the island's forests in two to make an international airstrip.

Easter Island: Much of Easter Island enjoys protected status. The island's airstrip has been recently extended to accommodate the American Space Shuttle.

Protected areas outside the South Pacific Region:

Hawaii: The US Fish and Wildlife Service has established an enviable system of parks and reserves here, the most notable being Haleakala on Maui, the Volcanoes National Park on the Big Island which contains some of the world's most active volcanoes, and to the west the Hawaiian Islands National Wildlife Refuge, which includes many reefs and small islands. These last are important for many threatened species, including the very rare Hawaiian monk seal *Monachus schauinslandi*. Koke'e State Park on Kauai is excellent for rare birds. The Kalalau trail here from Haena is breathtaking. The US Nature Conservancy has funded a number of reserves, including Hakalau Preserve on Big Island; Kipahulu Preserve and Waikamoi (notable for threatened populations of Maui *'akepa, Loxops coccinea ochracea*, crested honey-creeper *Palmeria dolei*, Maui parrotbill *Pseudonestor xanthophrys* and nene *Nesochen sandviciensis*) on Maui; Kamakou Preserve on Molokai; Kaluahonu Preserve on Kauai. All are outstanding for bird-watchers.

Johnston Atoll: An isolated atoll 1,320 kilometres south-west of Honolulu; also a US military base. Fifty-six species of bird have been recorded there, numbering 600,000 in the breeding season, including many migrating waterfowl. The monk seal was established there in 1960.

Galapagos National Park: Some 80 per cent of this remarkable archipelago is included within the Park, which is also listed as a World Heritage Site; 36 per cent of the species are endemic to the islands, including Darwin's finches, the famous giant tortoises and marine iguanas.

Cocos Island National Park: 670 kilometres off the Costa Rican coast. Three endemic birds including a cuckoo, flycatcher and finch. Large numbers of migratory sharks and dolphins pass through its waters.

Juan Fernandez National Park: Includes Robinson Crusoe, Alexander Selkirk, and Santa Clara Islands. 60 per cent of the plants are endemic, including cabbage trees *Dendroseris*, unique palms and tree ferns. A million Juan Fernandez fur seals, *Arctocephalus philippi*, were slaughtered each year at the end of the eighteenth century. Thought extinct in 1960, they were rediscovered in 1965 and now number several thousand.

BIBLIOGRAPHY

Adams, B. *The Flowering of the Pacific.* Collins, Sydney, 1986

Amesbury, S. S. and Myers, R. F. *Guide to the Coastal Resources of Guam* Vol. 1: *The Fishes.* University of Guam Press, Guam, 1982

Bagnis, R. and Christian, E. *Underwater Guide to Tahiti.* Editions du Pacifique, Papeete, 1985

Beaglehole, J. C. *The Exploration of the Pacific.* Stanford University Press, Stanford, 1966

Beehler, B. M., Pratt T. K. and Zimmerman, D. A. *Birds of New Guinea.* Princeton University Press, Princeton, 1986

Bellwood, P. *Man's Conquest of the Pacific.* Oxford University Press, Oxford, 1979

—— *The Polynesians: Prehistory of an Island People.* Thames and Hudson, London, 1987

Blashford-Snell, J. and Tweedy, A. *Operation Raleigh: Adventure Challenge.* Collins, London, 1988

Brosse, J. *Great Voyages of Exploration: The Golden Age of Discovery in the Pacific.* Doubleday, Australia, 1983

Carlquist, S. J. *Hawaii – A Natural History.* Pacific Tropical Botanical Garden, Honolulu, 1980

Celhay, J. cl., and Herman, B. *Plants and Flowers of Tahiti.* Editions du Pacific, Papeete, 1974

Cheesman, E. *Off the Beaten Track in the Far, Fair Society Islands.* Witherby, London, 1927

Clunie, F. *Birds of the Fiji Bush.* Fiji Museum, Suva, 1984

Danielsson, B. *Gauguin in the South Seas.* Doubleday, New York, 1966

—— *The Happy Island.* London Readers Union, London, 1954

Decker, R. and B. *Volcano Watching.* Tongg Publishing Co., Honolulu, 1984

Dousset, R. and Taillemite, E. *The Great Book of the Pacific.* Chartwell Books, New Jersey, 1979

Ellison, J. W. *Tusitala of the South Seas.* Hastings House, New York, 1953

L'Encyclopédie de la Polynésie. Diffusion Tahitienne du Livre, Papeete, 1987

Gould, J. *Population Pressures in the Pacific Islands.* South Pacific Peoples Foundation of Canada, Victoria, BC, 1987

Guppy, H. B. *Observations of a Naturalist in the Pacific between 1896–1899.* Macmillan, London, 1906

Hadden, D. *Birds of the North Solomons.* Wau Ecology Institute Handbook No. 8, Papua New Guinea, 1981

Haley, D. *Seabirds of the Eastern North Pacific and Arctic Waters.* Pacific Search Press, Seattle, 1984

Hannecart, F., and Letocart, Y. *Oiseaux de Nlle. Calédonie et des Loyautés.* Vols 1 & 2. Les Editions Cardinalis, Noumea, 1980 & 1983

Hargreaves, D. and B. *Tropical Trees of the Pacific.* Hargreaves Co. Inc., Kauilua, Hawaii, 1970

Heyerdahl, T. *Kon Tiki.* Rand McNally, Chicago, 1950

—— *Aku Aku.* Allen & Unwin, London, 1958

Hinton, A. G. *Shells of New Guinea and the Central Indo-Pacific.* Jacaranda Press, Sydney, 1972

Hobson, E. S. *Hawaiian Reef Animals.* University of Hawaii Press, Honolulu, 1972

Holyoak, D. T. *Guide to Cook Islands Birds.* D. T. Holyoak, Rarotonga, 1981

Johannes, R. E. *Words of the Lagoon: Fishing and Marine Lore in the Palau Region of Micronesia.* University of California Press, California, 1981

Kay, A. *Little Worlds of the Pacific.* Harold L. Lyon Lecture No. 9, Bishop Museum, Hawaii, 1980

Kent, G. *The Politics of Pacific Island Fisheries.* Westview Press, Colorado, 1980

King, W. B. *Seabirds of the Tropical Pacific Ocean.* Smithsonian Institute, Washington DC, 1967

Kirch, P. *The Evolution of the Polynesian Chiefdoms.* Cambridge University Press, Cambridge, 1984

Lewis, D. *We the Navigators.* University of Hawaii Press, Honolulu, 1972

Loveridge, A. *Reptiles of the Pacific World.* Macmillan, London, 1946

Lundsgaarde, H. P. *Land Tenure in the South Pacific.* University of Hawaii Press, Honolulu, 1980

McCoy, M. *Reptiles and Frogs of the Solomon Islands.* Wau Ecology Institute Publication No. 7, Papua New Guinea, 1980.

Macintyre, M. *The New Pacific.* Collins, London, 1985

Maude, H. E. *Of Islands and Men: Studies in Pacific History.* Oxford University Press, Melbourne, 1968

Mayer, J. F. *Birds of the South-West Pacific.* Macmillan, New York, 1945

Meek, A. S. *A Naturalist in Cannibal Land.* T. Fisher Unwin, London, 1913.

Merrill, E. *Plant Life of the Pacific World.* Charles E. Tuttle Co., Rutland VT, 1945

Mitchell, A. W. *The Enchanted Canopy: Secrets from the Rainforest Roof.* Collins, London, 1986

—— *Voyage of Discovery.* Severn House, London, 1982

Moorhead, A. *The Fatal Impact: The Invasion of the South Pacific 1767–1840.* Hamish Hamilton, London, 1987

Murray, J. *A Pattern of Islands.* Penguin, London, 1981

Muse, C. and S. *The Birds and Bird Lore of Samoa.* Pioneer Press, Washington, 1982

Pratt, H. D., Bruner, P. L. and Berrett, D. G. *The Birds of Hawaii and the Tropical Pacific.* Princeton University Press, Princeton, 1987.

Robin, B. *Living Corals.* Editions du Pacifique, Papeete, 1980

Ryan, P. *Fiji's Natural Heritage.* Southwestern Publishing Company Ltd., Auckland, 1988

Schmid, M. *Fleurs et Plantes de Nouvelle-Calédonie.* Editions du Pacifique, Papeete, 1981

Semper, K. *The Palau Islands in the Pacific Ocean.* F. Brockhaus, Leipzig, 1873. Reprinted by Micronesia Area Research Centre, Guam, 1982.

Snow, P. and Waine, S. *The People from the Horizon: An Illustrated History of the Europeans among the South Sea Islanders.* Phaidon Press, Oxford, 1979

Stanley, D. *Micronesia Handbook.* Moon Publications, Chico, California, 1985

—— *South Pacific Handbook.* Moon Publications, Chico, California, 1986

Stevenson, R. L. *In the South Seas.* Scribner's, New York, 1901

Suggs, R. C. *Hidden Worlds of Polynesia.* Cresset Press, London, 1963

Talbot, F. *Reader's Digest Book of the Great Barrier Reef.* Sydney, 1984

Thibault, J. cl., and Rives, cl. *Oiseaux de Tahiti.* Editions du Pacifique, Papeete, 1975

Watling, D. *Birds of Fiji, Tonga, and Samoa.* Millwood Press, Wellington, 1982

—— *Mai Veikau: Tales of Fijian Wildlife.* Shell Fiji Ltd, Suva, 1986

Whistler, W. A. *Coastal Flowers of the Pacific*. Oriental Publishing Co., Honolulu, 1980
Williams, T. *Fiji and the Fijians* Vol. 1. First published in London, 1858, reprinted by Fiji Museum, Suva, 1985
Wright, R. *On Fiji Islands*. Viking Penguin Inc., New York, 1986

Some Journals and Reports:
AMBIO. *The South Pacific: a Special Issue* Vol. 13, nos 5–6. Royal Swedish Academy of Sciences, 1984
Douglas, G. *A Checklist of Pacific Oceanic Islands*. Micronesia Vol. 5 (2), 1969.
Hay, R. *Bird Conservation in the Pacific Islands*. International Council for Bird Preservation Study Report No. 7, 1985.
Melville, R. *Lost Pacifica*. Nature, Vol. 211, London, 1966
Thaman, R. R. *The Poisoning of Paradise: Pesticides, People, Environmental Pollution and Increasing Dependency in the Pacific Islands*. South Pacific Forum, University of the South Pacific Sociological Society, 1 (2), Suva, 1984

INDEX

Page numbers in italics refer to illustration captions

Hawaiian Islands National Wildlife
  Refuge, 239
Hawaiian leaf-hopper (insect), 163
Hawaiian monk seal (*Monachus
  schauinslandi*), 240
hawksbill turtle (*Eretmochelys
  imbricata*), 88
head-hunting, 48; *see also*
  cannibalism
Hemignathus parvus see little honey
  creeper
Henderson Field, Guadalcanal, 35, 68
Henderson Island (Pitcairns), 239
Herald, HMS, 124
hermit crab, *176*
herring, 79
Heyerdahl, Thor, 26, 182, 191, 217,
  224–5, 227
hibiscus, 191, 199, 206
Hina (legendary mother of Maui), 149
Hiro (Tahitian god), 199
Holy Mama (of New Georgia), 57–8
honey-eaters (birds), 42, 52, 124, 196
Honiara, 34, 36, 87, 92
Honolulu, 156
hornbills, 54–5
Hotu Matua (ancestral King of Easter
  Island), 224
Huahine island (Leewards), 178,
  180–1, 192, 218
Humboldt current, 119
humpback whales, 96
hurricanes, 122–3, 172

Ice Age, 106–7
iguanas, 108–11, 115, 117–20, 240
i'iwi (bird), 169
Ile du Lys (Indian Ocean), 208
iliau (plant), 153
imperial pigeons, 104
Incas, 217
incubator birds *see* megapodes
Indians: in Fiji, 113–14
insects: number of species, 56, 132; in
  Hawaii, 161–4; *see also* individual
  species
International Union for the
  Conservation of Nature and
  Natural Resources (IUCN), 205
Iridomyrmex ants, 52
Iron Bottom Sound, 92
ivi tree, 129
Ixobrychus sinensis see yellow
  bittern

Japan: wartime campaigns, 66–9
Japan Trench, 20
Jaya, Mount (New Guinea), 107
Jersey Wildlife Preservation Trust,
  206
jet stream: carries insects, 161
Jim Boy (guide), 127, 129–32
Johnston Atoll, 240
Juan Fernandez fur seal
  (*Arctocephalus philippi*), 240
Juan Fernandez island, 123
Juan Fernandez National Park, 240
Jubea chiliensis see Chilean wine
  palm
jungle mynah (bird), 203
jungle nightjar, 123

Kadavu island, 129
kagu (*Rhynochetos jubatus*), 237
Kahoolawe island (Hawaii), 156
Kaluahonu Preserve (Kauai), 240
Kamakokahai (Hawaiian goddess),
  154
Kamakou Preserve (Molokai), 240
Kanakas (of New Caledonia), 212
Kaneshiro, Ken, 162–3
kapok flowers, 138
Karakakooa Bay (Hawaii), 157
Kau silversword (plant), 153
Kauai island (Hawaii), *80*, 156, 170–3,
  177
Kauai *kama'o* thrush, 173
Kauai *nukupu'u* (bird), 173
Kauai *o'o* (bird), 173
kava (Fijian drink), 117
Kawaikini peak (Kauai), 172
Keanawilinau (Hawaii), 154
Kelera (Fijian), 111, 115, 119
Kennedy, John F., 60
Kermadec Island, 237
Keti, Ahazi (Solomon Islander), 40–1,
  47
Kilauea (Hawaiian volcano), 155
King, Lieutenant James, 157, 182
kingfishers, 124, 209, 211
Kipahulu valley (Maui), 166, 169–70
Kipahulu Preserve (Maui), 240
Kiribati, 18, 234, 238
koa tree, 165, 169
Koke'e State Park (Kauai), 172, 240
Kolombangara (volcano), 51, 57–8
Kolombangara Forest Reserve
  (Solomons), 237
Kolovai village (Tonga), 146–8
Koro island, 127